How To Build Brackets

By Greg Vanden Berge

Published by Greg Vanden Berge
Copyright 2016 Greg Vanden Berge

ISBN-13: 978-1532781889
ISBN-10: 1532781881

Stair Building Books

Advanced Stair Stringer Layout Methods
Professional Stairway Building Secrets
How To Build And Frame Stairs - Book 1
How To Build And Frame Stairs With Landings - Book 2
How To Build And Frame Winder Stairs - Book 3

Greg's Other Construction Books

501 Contractor Tips
Simplified Tile Floor Installation
Simplified House Inspection Checklist
Simplified Home Inspections
Guide For Hiring Contractors
Home Buyers Checklist

http://gregvandenberge.com

http://www.homebuildingandrepairs.com

Disclaimer

Greg Vanden Berge, and its owners, agents and employees, make no warranty respecting the accuracy or currency of any information in the content or pages of this book or any source document referenced herein or linked to herein. Use of this book is conditioned on the user's understanding and agreement that we shall not be liable, on any theory whatsoever, including but not limited to negligence, for any damages attributable to that use.

In no event shall Greg Vanden Berge, its owners, agents or employees be liable to you or anyone else for any decision made or action taken by in reliance on any content created by Greg Vanden Berge or other individuals, companies, corporations or parties.

Greg Vanden Berge and its affiliates, agents, owners and employees shall not be liable to you or anyone else for any damages, including without limitation, consequential, special, incidental, indirect, or similar damages, even if advised of the possibility of such damages.

Your use of this book and all related rights and obligations, shall be governed by the laws of the United States of America, as if your use was a contract wholly entered into and wholly performed within the United States of America.

Any legal action or proceeding with respect to this book or any matter related thereto may be brought exclusively in the courts of the United States of America. By using this book, you agree generally and unconditionally to the jurisdiction of the aforesaid courts and irrevocably waive any objection to such jurisdiction and venue.

Do not copy or distribute this book. This manual contains materials protected under International and Federal Copyright laws and Treaties. Any unauthorized reprint or use of this material is prohibited.

Table of Contents

Introduction	4
What is a Stairway	5
What is a Bracket Stairway	6
Tools	7
Hardware and Fasteners	8
Stair Parts	9
Safety	10
Engineering	11
Calculating Riser Height	14
Calculating Tread Run	17
17.5 Tread and Riser Rule	19
Layout and Measurements	21
Stair Stringer Layout	23
Building The Stairs	58
Glossary	72
Rise and Run Chart for 10 Inch Treads	80
Rise and Run Chart for 17-½ Rule	94
Decimals to Inches Chart	108
Bracket Stair Building Examples	109

Introduction

This if Book 5 - How to Build and Frame Stairs with Brackets

This book provides you with step-by-step detailed instructions on how to design, layout and build a variety of different stairs using tread brackets. Sections of this book were copied from Book 1 - How To Build And Frame Stairs and Simplified Bracket Stair Building.

This book is part of a series designed for professionals and do-it-yourselfers to provide them with what I consider to be a simplified step-by-step process for designing and assembling different types of stairs. Each book will be written and illustrated specifically for the type of stairway specified in the title.

Book 1 - How To Build And Frame Stairs

Book 2 - How To Build And Frame Stairs With Landings

Book 3 - How To Build And Frame Winder Stairs

Book 5 - How to Build And Frame Stairs With Brackets

We are currently working on the next books listed below.

Book 4 - How to Build Circular or Curved Stairs

Book 6 - How to Build Stairs with Odd Shapes

Book 7 - How to Build Stairs with Dado Stairways

Book 8 - How to Build Stair Handrails

The author has built over 1000 stairways and taught people in different parts of the world how to build a variety of different types of stairways. He has spent 35 years in the construction industry working on practically every type of project you can imagine as a master carpenter.

What is a Stairway?

A stairway or set of stairs is usually a series of steps used to take you from a lower level to an upper-level.

Before

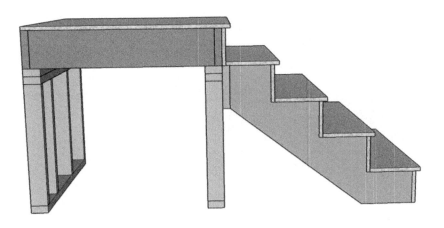

After

What is a Bracket Stairway?

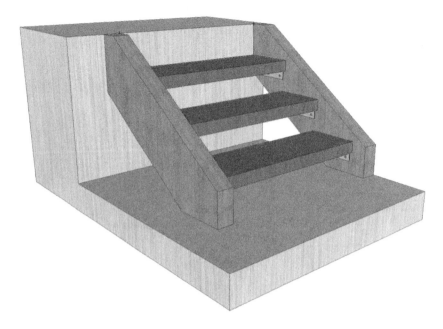

A bracket stairway uses metal brackets like the one shown below to attach treads or steps to stringers.

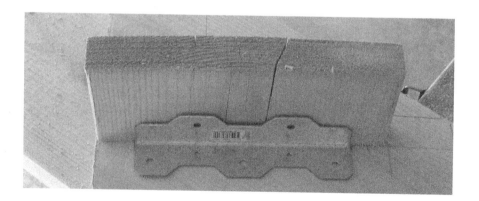

Tools

Hammer

Level

Pencil

Circular Saw

Framing Square

Tape Measure

Caulk Line

Optional Tools – nail puller, ladder, carpenter pouch or nail bags.

Hardware and Fasteners

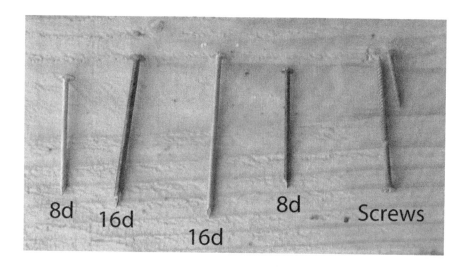

Recommend using 16d nails for materials between an inch and one quarter (1 1/4") and an inch and a half (1 1/2").

Recommend using 8d nails for materials between one half-inch (1/2") and three quarters of an inch (3/4").

I can't recommend screw sizes, because some building departments and engineers don't recommend them to be used for assembling stairs using construction standard lumber.

I also recommend using galvanized or stainless steel nails for building "exterior" stairs out of wood.

For additional holding power you can use ring shank, hot dipped or drive screw deformed shank type nails for stair treads and sheathing.

Stair Parts

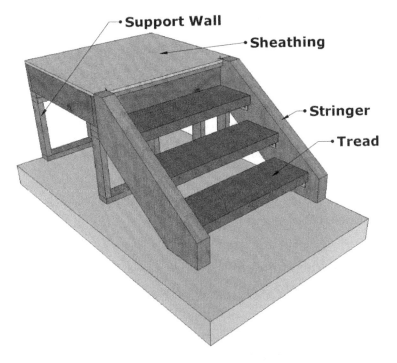

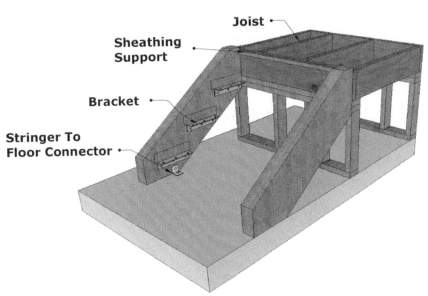

Safety

This book does not provide you with detailed safety information and recommends gathering more information on tool use, demolition, safety and assembling practices.

Safety Recommendation Tips

1. Avoid standing on stair stringers, joist and other framing members that haven't been properly secured.

2. Avoid using power tools near water and make sure they are in good condition.

3. Use personal protection equipment like safety glasses, slip resistant shoes, earplugs, hardhats and gloves when necessary.

4. Inspect all ladders for defects before using them and make sure you use the appropriate ladder for each phase of the project.

5. Use extra caution when working with any materials, including prefinished materials like stair treads that are or can become slippery.

6. Keep the job site and working areas clean.

7. Be prepared for emergencies with an on-site first aid kit and directions to the nearest hospital or emergency care facility.

8. Always use the right tool for the right job.

9. It's a good idea to create a barricade around the area you're going to be working in, especially when working on your own home around pets and children.

10. Avoid lifting heavy objects and work smart.

Engineering

I won't be able to provide you with exact lumber sizes for your particular project, but can provide you with a few recommended lumber sizes from projects I've built in the past that were approved by building professional's.

Again, these are only recommended construction standard lumber sizes and might not work on your project.

Estimated Stair Stringer Sizes

3 x 12 for lengths up to 8 feet.

4 x 12 for lengths up to 18 feet.

You would need to check with the manufacturer for stringer sizes if you're planning on using engineered lumber.

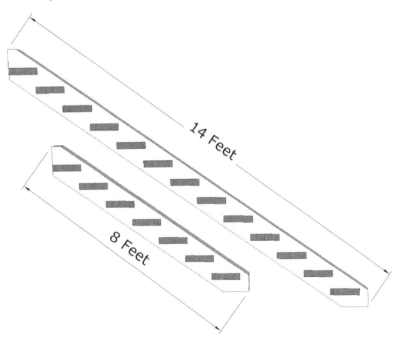

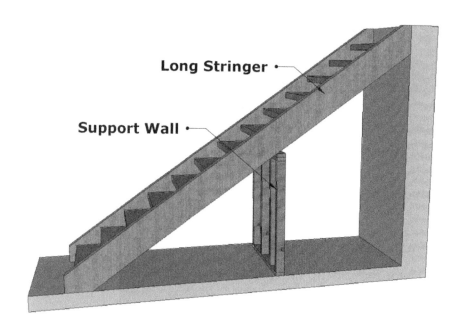

Stair stringers longer than 10 feet might require supporting walls. 2 x 4 can usually be used to build supporting walls less than 10 feet high for stringers, landings and floors.

Landing or Platform Joist

2 x 6 joists usually works for most landings with joist spans less than 6 feet and spaced 16 inches on center.

2 x 8 joists usually works for most landings with joist spans less than 8 feet and spaced 16 inches on center.

2 x 10 joists usually works for most landings with joist spans less than 12 feet and spaced 16 inches on center.

Use metal building hardware or treated lumber framing base plates when attaching a wood framed stairway to concrete.

Stair Treads and Stringer Spacing

Stair treads and risers can be made out of engineered lumber like plywood and oriented strand board or construction standard lumber.

Estimated stringer spacing for 2 x 12 construction standard nominal sized lumber used for treads would be no more than 24 inches.

Estimated stringer spacing for 3 x 12 construction standard nominal sized lumbers used for treads would be no more than 36 inches.

Estimated stringer spacing for 4 x 12 construction standard nominal sized lumbers used for treads would be no more than 48 inches.

The lumber sizes I've just provided you with are fairly conservative and should work with most engineering load requirements of 40 pounds per square foot or less which is usually acceptable for residential stairways.

These are only estimates and might not work for stairways designed for public, commercial or industrial type buildings. These buildings usually require larger lumber sizes to handle extra weight from additional people and large items.

You can contact an engineer for more information.

Calculating Riser Height

If you have a set of building plans then you should already have the correct individual tread and riser measurements. This chapter and the one that follows will be more helpful to those who will be designing their own stairs.

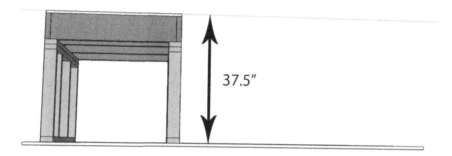

The first thing we need to do is measure the total stair rise which is the measurement between the lower and upper level. In our example we will be using 37 ½ inches and will provide you with a simple and easy formula for figuring out your individual stair riser measurement in the next two pages.

The next thing we need to do is figure out how many stair steps (treads) and risers we will need for the stairway and the easiest way I can think of doing this will be to divide the stair rise by the number seven.

Example: 37.5 ÷ 7 = 5.36 and then we will round the number off to its nearest whole number (5). The number five will provide us with a good place to start, but it won't be a bad idea to divide the next smaller and larger whole numbers into the total rise to find the most comfortable individual riser measurement.

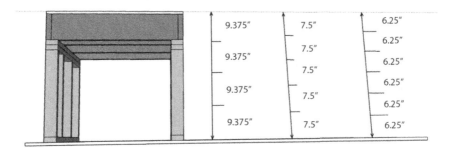

In the example above we will divide 4 (the next smaller whole number), 5 and 6 (the next larger whole number) into the total rise of 37.5 inches.

37.5 ÷ 4 = 9.375"   37.5 ÷ 5 = 7.5"   37.5 ÷ 6 = 6.25"

It provides us with two choices, because some building codes don't allow risers to be taller than 7 ¾" or smaller than 4 inches, leaving us to choose from either a 7.5 or 6.25 inch measurement.

Either one is acceptable, but before making your final decision, I suggest reading the next two chapters especially the one on the 17 ½ inch rule.

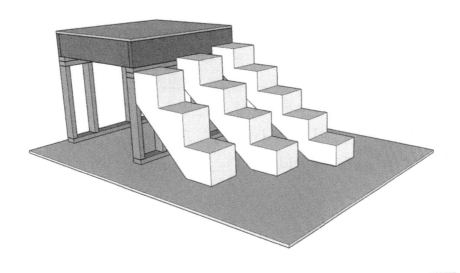

Calculating Riser Height Formula

Using Previous Example

Total Rise in Meters or Inches _ 37.5

Total Rise _ 37.5   Divided by Seven Equals _ 5.36

Rounded off Whole Number (B) _ 5

Next Smallest Number (A)   _ 4

Next Largest Number (C)   _ 6

Total Rise_37.5   Divided (A) _4   Equals _9.375

Total Rise_37.5   Divided (A) _5   Equals _7.5

Total Rise_37.5   Divided (A) _6   Equals _6.25

You Can Use Either Meters or Inches in Formula Below.

Total Rise __91¼__

Total Rise __91¼__ Divided by Seven Equals __13.035__

Rounded off Whole Number (B) __7.01__

Next Smallest Number (A) __7.6__

Next Largest Number (C) __6.51__

Total Rise __91.25__ Divided (A) __7.6__ Equals __12__

Total Rise __91.25__ Divided (B) __7.01__ Equals __13__

Total Rise __91.25__ Divided (C) __6.51__ Equals __14__

Calculating Tread Run

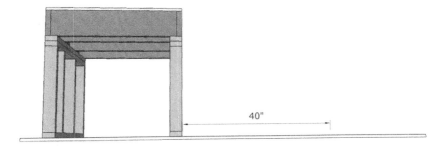

We will use the same method we used to calculate our risers. Simply take the measurement you're going to use for the overall tread run and divide it into equal measurements.

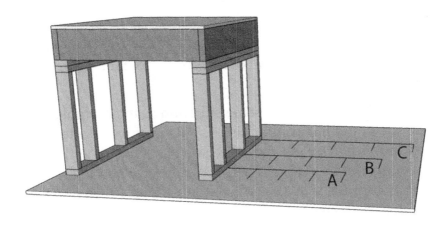

A - Tread Run = 32 inches and four 8 inch treads.

B - Tread Run = 40 inches and four 10 inch treads.

C - Tread Run = 48 inches and four 12 inch treads.

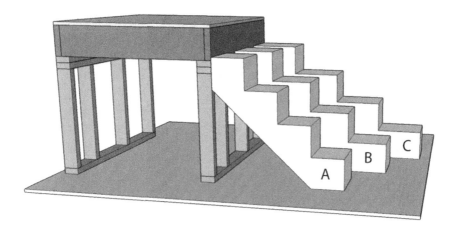

There is always going to be one more riser than treads, so you can use this to figure out either the amount of treads or risers you'll be using for your stairway.

I suggest figuring the amount of risers that will be used and then subtracting one from the total amount to calculate the amount of treads.

For Example: If you have 14 individual 7 ¾" risers then you will have 13 individual treads. It's hard to do it the other way around.

For Example: If we use the same situation as the example above we would have a total rise of 108 ½". If we decided to use 11 individual treads and added one to come up with the total number of individual risers it would give us 12. If we divide 12 into 108 ½" we would end up with a 9 inch individual riser height.

There will be times when you need to start with the total tread run, but most of the time starting with the total rise will provide you with the best way to figure out the total run, amount of treads and individual tread width.

17.5 Tread and Riser Rule

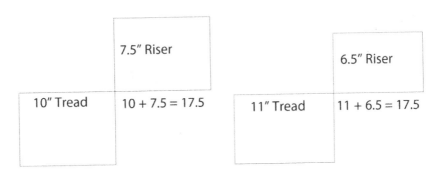

Here's a method I've been using for years to provide the most comfortable step possible, even though you might not be able to use it all the time, because it won't work in confined areas.

Simply add the individual tread and riser measurements together as shown in illustrations above. The closer you come to 17 ½ inches the better.

In the example above we are way off with a 16 inch tread and 7 ½ inch riser, but the solution to the problem wouldn't be to use a 1 inch riser. It would be better to use a smaller tread or a smaller riser as long as the total individual riser height isn't less than 4 inches.

Remember, some residential building codes discourage using risers that have an individual height of more than 7 3/4" or a minimum height less than 4 inches and I agree.

The example above also doesn't provide us with something reasonable, even though the numbers add up to 17 ½ inches.

I don't suggest using individual stair tread widths less than 10 inches and some building codes don't recommend using treads less than 11 inches. As you can see by the illustration in the upper right-hand corner, people with larger shoes might have problems using stairs with smaller tread widths.

Remember, you don't need to build a stairway using this formula, but it does provide us with a comfortable step if used within reason.

Listed below are a few examples of what I would consider to be comfortable steps.

10 inch stair tread width + 7 to 7-3/4" riser height.

10-1/2 inch stair tread width + 6-1/2 to 7-1/4" riser height.

11 inch stair tread width + 6 to 6-3/4" riser height.

11-1/2 inch stair tread width + 5-1/2 to 6-1/4" riser height.

Even though 17 ½ inches will provide us with a comfortable step, anything between 17 and 18 inches should be acceptable. Anything between 16 ½ and 18 ½ inches could be acceptable also, but once you start to get out of this range, then steps might start to get uncomfortable.

Layout and Measurements

By now you should have figured out the overall height (total rise) and the overall length (total run) along with the individual riser height and tread width. However, there are a few things I would like you to check before laying out your stair stringer.

1. Is there anything you need to add or subtract to the total rise or run and if so then you will need to recalculate to find the new individual tread or riser measurements?

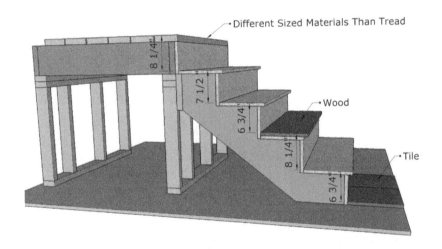

If you're going to install different sized materials on either the stairway steps or upper and lower levels like floors or landings then you will need to make the necessary adjustments while laying out the stair stringer.

2. Is it going to fit into the designated area?

Now would be a good time to double check all of the stairway measurements by re-measuring the total rise and total run and verifying that you have done all calculations correctly.

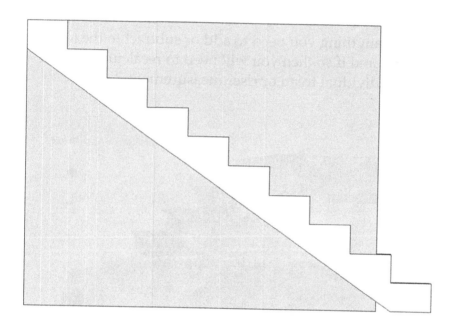

3. Is everything square and parallel?

The area where the stairway is going to be installed is usually called the stairwell and parallel walls will need to have the same measurement at both ends and in the middle. 90° perpendicular walls will need to be square or as square as possible in the stairwell.

You should also check to make sure all connecting walls are level and plumb. It will be easier to make the necessary adjustments before the stairway is installed.

Stair Stringer Layout

I've simplified the stair building process by providing you with illustrations, pictures and now we're going to throw in a few videos. If a picture is worth a thousand words then a video must be worth a couple hundred thousand or even a million words.

I've made plenty of them and if you have access to the Internet, should visit our website, to make your stair building process even easier.

Laying out the top and bottom of the stair stringer will be the most difficult part of this process. Hopefully by now you have enough information and figured out your individual riser heights and tread widths, along with the amount needed, to build the stairway.

In our example we're going to be building a set of stairs with three steps up to a platform.

The methods for constructing our sample stairway can be applied to almost any type of stair building situation. If you wanted to build a five step set of stairs, all you'll need to do is continue laying out two more steps.

The basic fundamentals for laying out one stair stringer can be used to lay out others and it really won't matter how big or small the stairway actually is.

For our sample set of stairs, we'll use a 30 inch total rise and a 30 inch total run. We will be using 10 inch wide stair treads and 7 1/2 inch stair risers.

Setting Up The Framing Square

In the picture above, I've already set up the framing square. We're not going to use the outside of the framing square, we're going to use the inside (the outside of the framing square is where we'll be attaching the stair gauges).

You can put the stair gauges on the inside of the framing square, but I usually don't, because they'll be in the way, when marking stair treads and risers.

If you don't have any stair gauges you'll need to adjust the framing square for each stair riser and tread, before marking the lumber. In the next two pictures, I will zoom in and show you how to position the framing square and use the stair gauges.

10 Inch Tread Mark On Framing Square

In the picture above the arrow is pointing to a line that represents the 10 inch mark on the framing square. Hopefully, you can see the number 10. The marking line runs in between the one and the zero, of the number 10.

We want to line up this line with the edge of our stair stringer material (2 x 12).

This side of the framing square will represent your stair tread width and if you're using an 11 inch stair tread, make sure you line up the number 11 on your framing square with the edge of your stair stringer material.

7 1/2 Inch Mark On Framing Square

In the picture above, the arrow is pointing to the line which represents 7 1/2 inches on the framing square. You can clearly see the seven and 8 inch mark on this side of the framing square.

You'll be lining up the 7 1/2 inch mark on the framing square with the edge of the stair stringer material. Once you have your framing square in position, you can start marking stair treads and risers.

This side of the framing square will represent your stair riser height and if you have a 7 1/4 inch stair riser then you would line the 7 1/4 inch mark on your framing square up with the edge of the material you're going to use for your stair stringer.

Marking The Stair Stringer

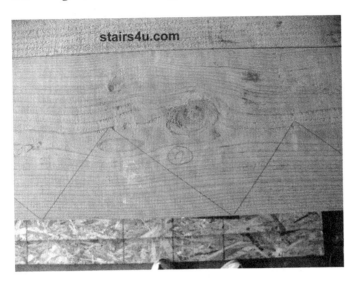

1. Once you have the framing square in the correct position, where the 7 1/2 inch mark on the framing square and the 10 inch mark are located at the edge of the lumber. Grab your pencil and mark your first step.

2. Then slide the framing square to your left until the 10 inch mark on the framing square lines up with the previous mark you've already made for the stair riser.

If you're using stair gauges, you'll simply continue to slide the framing square to the left, marking as many stair treads and risers as you need to.

3. Once you've lined up the 10 inch stair tread mark on the framing square with the riser mark you have already made from step one and have positioned the framing square correctly, feel free to grab your pencil and mark your second step.

4. Slide the framing square to your left and repeat step number three, until you have three stair treads and risers marked out on the lumber you'll be using.

In the picture above we marked out three tread and riser sections using our framing square. In the next steps I will show you how to lay out the bottom and top of your stair stringer.

Notes

Laying Out Stair Treads

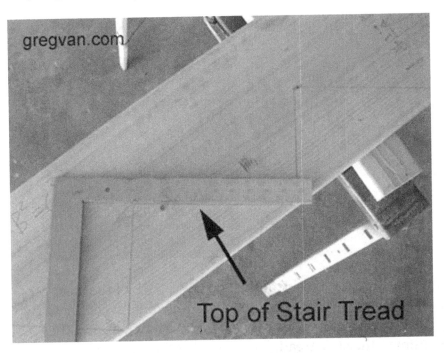

After you layout the stair treads and risers, you will need to mark the exact position of the stair tread. In this example we're using 1 1/2 inch thick, 2 x 12 stair treads.

We are going to use the 1 1/2 inch side of the framing square, to keep things simple. You could always cut an extra scrap piece from the stair treads you're going to use, if that works for you. Remember, we're trying to keep it as simple as we can.

For Example: If you're going to use 3 x 12's for your stair treads, instead of marking 1 1/2 inch stair tread widths, you will need to make them 2 1/2 inches. Simply replace the 1 1/2 inch mark with a 2 1/2 inch mark, so that your stair tread brackets can be positioned correctly.

Laying Out Stringers For 3 X 12 Treads

(Skip this section of the book if you're using 2 x 12 stair treads.)

Make a 2 1/2 inch mark at the front and back of the stair tread. Then you can draw a straight line in between these two marks, marking the top of the stair tread brackets and the bottom of the stair tread.

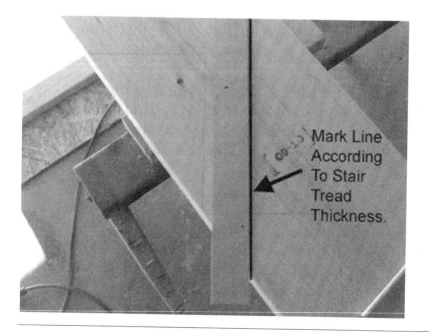

Marking Top Of Stair Tread Brackets

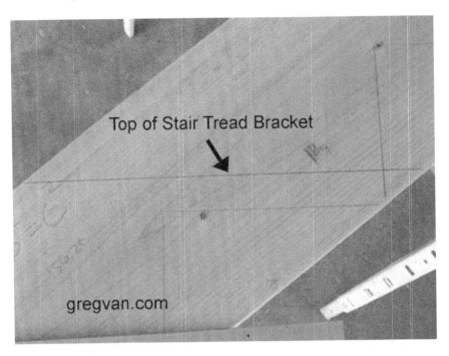

Your stair treads sit on top of the stair tread brackets.

Get Extra Help And Videos At Our Website:

http://www.homebuildingandrepairs.com/stairs/index.html

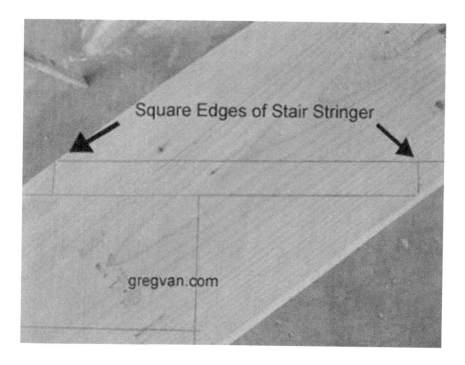

After you've marked the bottom of the stair tread, on the stair stringer, extend the top tread line, making sure you have two continuous lines.

The next step will be to square the corners at the edge of the stair stringer. This will provide us with a way to locate the center of the stair tread. Making you look like a professional stair builder. I can't tell you how many times I've ran across a stairway where the treads were located incorrectly.

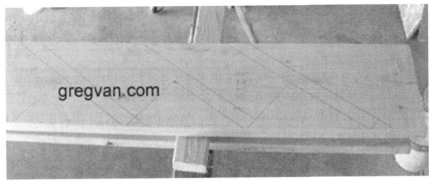

This is what your stair stringer should look like at this point of the project. If you notice, I only marked the bottom stair tread at both ends. We only need to mark the stringer corners on one stair tread.

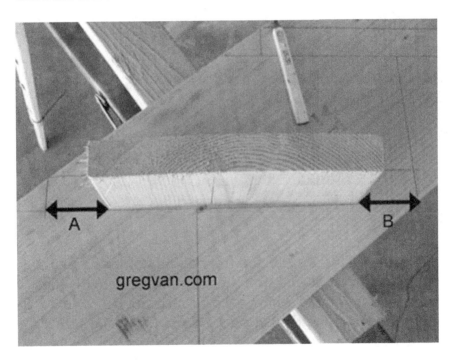

The next step will be to cut yourself a piece of scrap stair tread. Place it in between the top and bottom stair tread marks on the stair stringer, exactly like the picture above.

Then center the stair tread in between the stair stringer corner marks.

After you've located the center of your 2 x 12 stair treads, you need to mark the front and back of the stair tread. Once we have the front tread measurement, we can use it to mark the locations of the others.

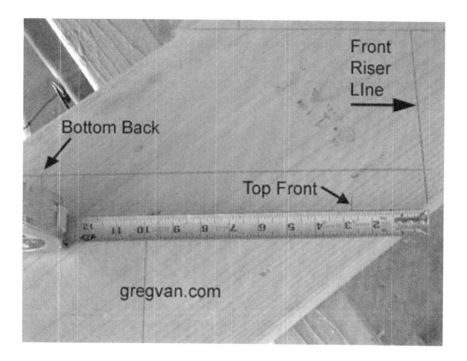

I made a small mark at the top front of the stair tread and the bottom back. Measuring from the front riser line to the top front of the stair tread, I have a 2 3/4" measurement. Next you will need to measure the bottom back mark to make sure it's 2 3/4" also.

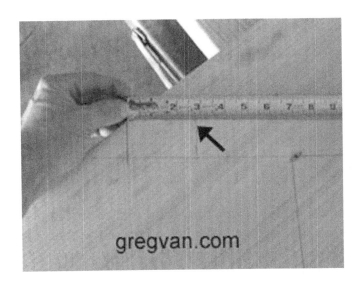

Double check everything, when laying out your stair stringers. Any mistakes you make here can be corrected with little effort. This might not be the case a little further in the project.

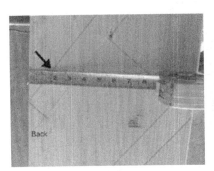 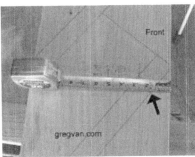

You can also measure from the edge of the stair stringer to the stair tread marks, for additional verification. Simply measure the front and back of the stair stringer at the points shown in the pictures above.

These measurements should be the same or within ¼ of an inch, unless you're a perfectionist. If there's a big difference, now's the time to make the necessary adjustments and figure out what you did wrong and correct it.

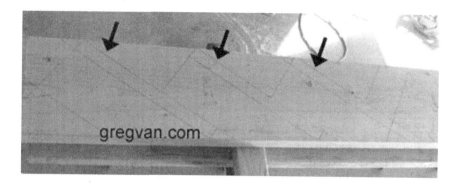

After you've double checked everything and you're absolutely positive the stair treads are located in the center of the stair stringer, mark the rest of the 2 x 12 stair tread locations on your pattern. These marks will be used for the front of the stair treads only. There's no need to mark the back of every single tread.

Laying Out Bottom Of Stair Stringer

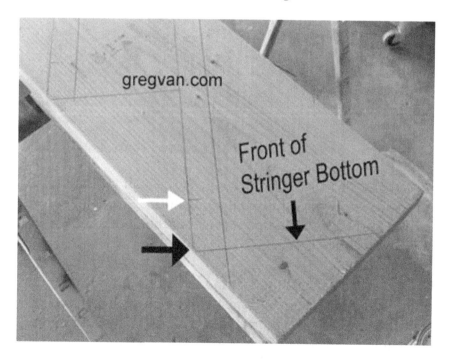

In the next step we will mark the front line of the bottom of the stringer. You can adjust this line if needed, depending upon your handrailing or stairway design. I usually go off of the original layout of the front riser and tread corner. The white arrow is pointing to the front edge of the stair tread location.

Here's what the finished front of the stair stringer bottom will look like. The black arrow is pointing to the front of stringer bottom and the white arrow is pointing to the front stair tread mark on the finished stair stringer.

Don't get confused, the finished stair stringer in the picture above is the stringer on the right side of the stairway, where the unfinished stair stringer we're currently working with is the left one.

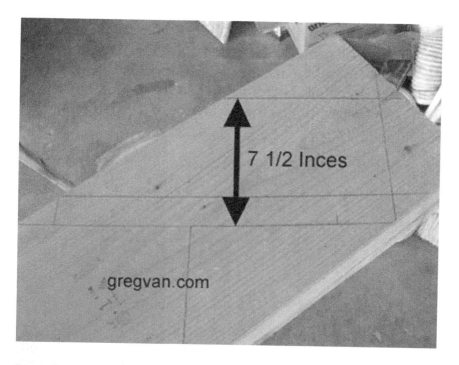

In order to get the next mark and complete the bottom stair stringer layout, we need to make a parallel mark with the bottom stair tread. This mark will be 7 1/2 inches or the distance of one of our stairway risers.

If your stair risers are different, make the necessary adjustments. The bottom stringer layout above is for a stair stringer that is going to sit directly on top of the existing floor.

For Example: If your stair risers are 7 inches each, you will need to replace the 7 1/2 inch measurement in the picture above with a 7 inch measurement. Remember, our stairway measurements might not be the same as yours and you will need to make the necessary adjustments, while you're laying out your stair stringers.

Laying Out Top Of Stair Stringer

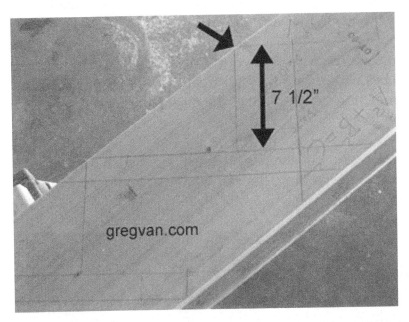

To find the top of the stair stringer, simply measure the distance of your stair riser, which in our case is 7 - 1/2 inches, from the top of the stair tread. Again, this measurement will need to be adjusted, if you have a different riser measurement.

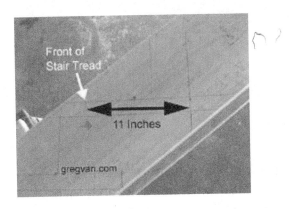

To get the stair stringer head out connection mark, simply add 1 inch to your tread run measurement.

Our stair tread run is 10 inches, so if we add 1 inch to that, we will have an 11 inch step, while keeping the 10 inch tread width we're using to layout the stringers.

Don't confuse the widths of your stair treads (2 x 12's that are 11 ½ inches wide), with the stair tread run (10 inches). This is a big problem for even some of the most experienced stair builders.

The 1 inch represents the nosing overhang, which could create a problem at the landing, floor or whatever you're attaching the stringers to.

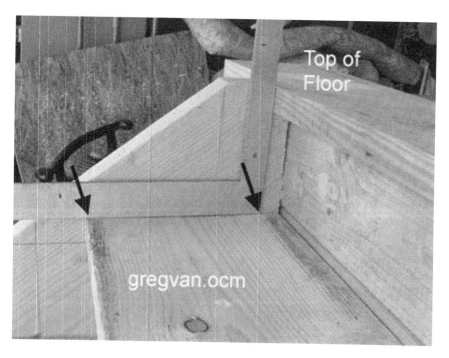

The distance in between the black arrows represent the individual 10 inch tread run. Without the additional inch, it could create an uncomfortable stairway and might not pass local building inspections.

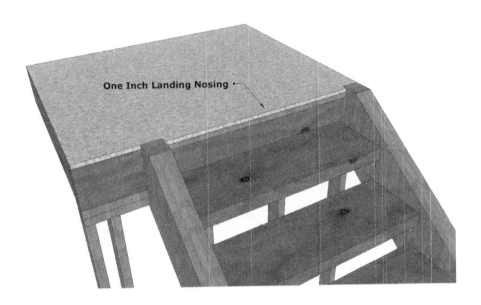

Let's see if these illustrations provide you with a better understanding of the problems you could create if the stringers aren't laid out correctly. Remember, the landing nosing isn't always going to be 1 inch and might need to be modified, depending upon your tread width and overhang measurements.

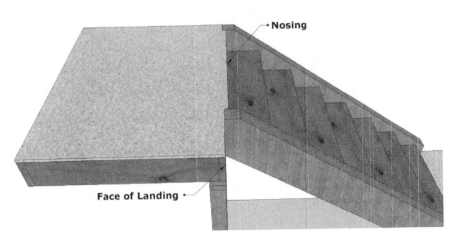

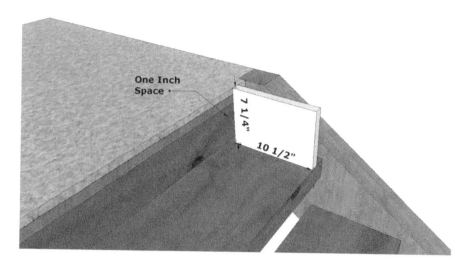

In these illustrations we're using a 7 ¼ inch riser and 10 ½ inch wide tread. The rectangle will represent each step and in both the top and bottom illustrations the top and front side lines up perfectly with the front corner of each step.

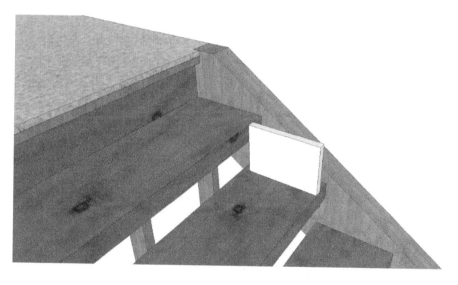

You should be able to place this rectangle on any step with similar results, if not, then you might have a problem with your treads, risers and stringer layout.

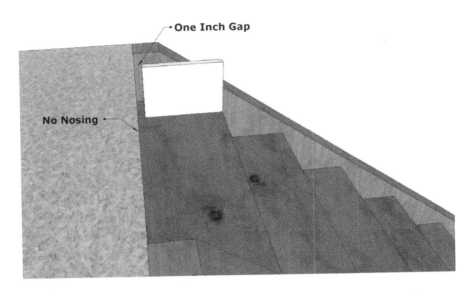

Without the nosing you can clearly see the problem and that is the 1 inch gap between the top corner of the landing and the rectangle we're using as a template to represent the tread and riser measurements.

Completed Stair Stringer Layout

Here's a picture of the completed left side stair stringer.

Cutting The Stair Stringer

For this particular stair stringer, you will only make four cuts. Each one of the arrows points to the individual lines that will need to be cut.

After the stair stringer is cut, we will place it into position to verify that it has been cut and laid out correctly. There's nothing worse than cutting all of your stair stringers, only to find out later that you made a mistake.

Place the stair stringer into its proper position and double check your measurements, before laying out and cutting the next stringer.

Make sure you have the correct stair riser measurement at the top and bottom connection points. In this case our stair riser is 7 1/2 inches. Simply hook the top of the landing with the tape measure and measure to the distance in between the top of your stair tread and top of the landing.

Notes
_____
_____
_____
_____
_____
_____

In this picture I'm checking the bottom stair riser height. If you look at the tape measure close enough, you will notice that it's measuring about 7 5/8". You can make the necessary adjustment by trimming an eighth of an inch off the bottom of the stair stringer.

An adjustment like this is going to be easier to make, before you attach the stair stringers, to the stair landing or floor. However, if you just left it alone and didn't make any adjustments it would be extremely difficult for anyone to ever notice that the lower riser was an eighth of an inch higher than the other ones. 1/8 of an inch variation in between risers and treads are perfectly acceptable in rough stair framing.

Laying Out The Other Stringer

For conventional stairs we can use one of our stair stringers as a pattern, but for bracket stairs, we will need to lay out both of the stair stringers individually.

However, I don't need to go over the entire process again, use the same simple steps you used for laying out the left stair stringer, for the right one. The only difference is that you will need to lay the stair stringer out in the opposite direction.

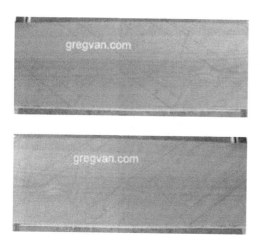

They need to oppose each other. In other words, the inside of the right stair stringer would be the outside of the left stair stringer and vice versa.

Here's a good picture of the finished right and left stair stringers. In this picture you can clearly see the difference between the two stair stringers.

Attaching Tread Brackets To Stringers

By now you should have a pretty good idea, of what you're doing. The stair tread bracket is going to attach to the stair stringer, using lag screws. If you look closely at the picture, you will notice that we centered the stair tread bracket in between the front and rear of the stair tread. This is one method, but I will show you a better one in the following steps.

Notes

The stair tread angle bracket, in the picture above is still positioned in the center of the stair tread, but I'm pointing to another pencil mark I made. I'm going to move the stair tread bracket a half inch forward.

This will reduce the possibility of the front of the 2 x 12 stair tread splitting off, simply because it wasn't supported properly. They sell longer brackets and I would recommend using them, if you can find them. The only reason why I used the 8 1/4 inch stair brackets was because my local home improvement center didn't carry the 10 1/4 inch brackets.

Since I'm not a structural engineer and this book only provides you with stair building basics. You should check with a local engineer or building contractor for more information about stairway structural building components, including stair stringers sizing, lumber dimensions for other stair parts and tread bracket sizes.

Remember, this book only provides you with basic stair building information. The dimensions and sizes of building materials will change, depending upon each different individual stairway use and dimensions.

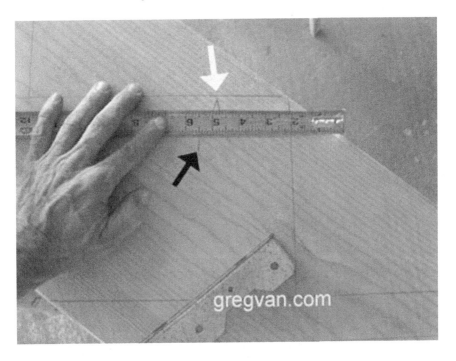

The next step will be to measure and mark the stair tread bracket location for each stair tread. The black arrow is pointing to the pencil mark that will be used to position the front of the stair tread angle bracket. The white arrow is pointing to the pencil mark that will be used to locate the front of the stair tread.

Next we will place each stair tread bracket into its proper position.

Notes

Left Stair Stringer

After you've placed the stair tread brackets in the correct position, double check to make sure you're installing them in the correct positions.

Right Stair Stringer

Just make sure you're not going to fasten the stair tread brackets in the wrong spot. Below is a helpful tip that I use often to remind myself about the position of the stair treads. These pencil marks can be lifesavers.

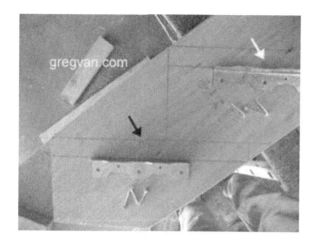

The black arrow is pointing to the word top, reminding the stair builder that this is the top of the stair tread. The white arrow is pointing to the letter B, representing the bottom of the stair tread.

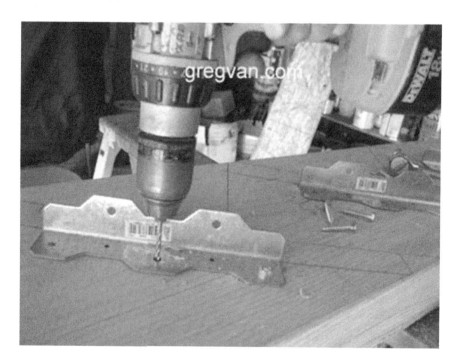

The next step will be to pre drill a few holes and then use the lag screws to attach the stair tread angle brackets to the stair stringers. Make sure you use a smaller drill bit if you're going to pre-drill any holes.

Example: The lag screws we're using are a 1/4" inch in diameter and 1 - 1/2 inches long. I don't recommend using a drill bit larger than 1/8" in diameter to pre-drill your holes, if you're using quarter inch diameter lag screws.

Remember, this is a sample stair project and you should always follow the manufacturer's installation instructions, if they're available.

Some stair tread bracket manufacturers require you to use a specific type of screw or bolt. If you don't follow the manufacturer's installation instructions, you could be putting yourself and others at risk.

You can use a wrench to tighten the lag screws or a socket and ratcheting device, like the picture above. Make sure each one of these screws is tight, but don't over tighten them.
If you apply too much pressure, while tightening these bolts, the head could break off, creating problems and more work.

The finished bracket should look something like this.

Notes

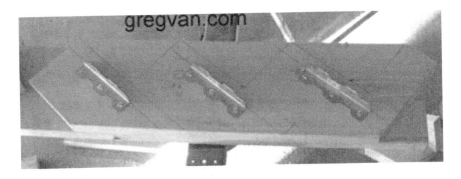

Here's a picture of the finished left stair stringer.

Here's a picture of the finished right stair stringer.

All we have left to do is set the stringers in place, cut the individual stair treads and attach them to the brackets.

Notes
_____
_____
_____
_____
_____
_____
_____
_____
_____
_____

Building The Stairs

Stair Stringer Attachment

The next step will be to attach the stair stringers to the floor and stair landing. In this step, you will be placing them in their permanent position.

Line the top of the stair stringer up with the top of the deck, landing or finished floor. When you have it in the correct position, you can use nails or screws to attach the stair stringer to the upper floor area.

I placed three black dots on the stair stringer. These will represent the areas you need to toenail the stair stringer to the upper floor area. Use at least 4 – 16d nails and galvanized nails or screws for exterior stairs.

Nailing Stair Stringers

Even though this isn't the same stair stringer, you can use the same method to attach your stair stringers to a ledger or rough floor framing.

_____
_____
_____
_____
_____

Don't forget to properly position and nail the bottom of the stair stringer to your wood or concrete floor. If you're nailing into a wood floor or deck, you can use the same method for attaching the stair stringer to the upper deck.

Simply toenail four 16d nails at an angle through the stair stringer into the wood floor. You can always use other home building hardware to strengthen your upper and lower stair stringer attachments to wood framing.

Here's a couple of heavy-duty angle brackets that are even located in a good spot. If you're going to attach your stair stringers to a concrete floor, make sure you use approved building hardware and fasteners.

You can use a variety of different fasteners including lag screws, nuts and bolts, concrete epoxies and anchors to connect the stringer to a floor or landing.

If you're building a large stairway you can contact a structural engineer for more information on how to properly connect your stair stringers to the top and bottom of the building.

Whatever you do, make sure you locate these brackets far enough back so that people won't be stubbing their toes on them. If these brackets had been moved more than 3 inches forward, it could have created a safety hazard.

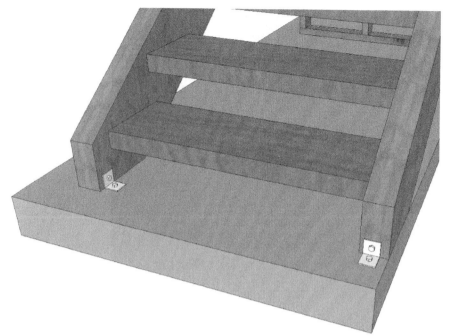

The upper illustration provides us with bad locations and the bottom illustration provides us with good ones.

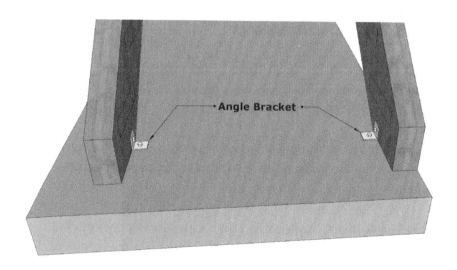

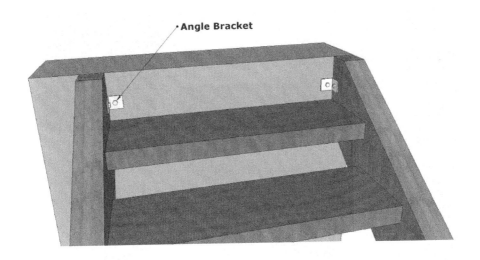

Even though there are other methods used to connect the upper section of a stair stringer to a floor, these would be the most common. You can also use larger angle brackets, these have one hole on each side, but you can get them with two or three holes on each side also.

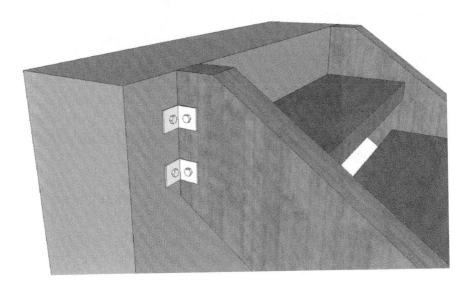

After you have attached the right stringer, you can attach the left one. Don't forget, this is only a sample stairway that's 2 feet wide. The minimum width for most stairways is 36 inches. Check with your local Building and Safety Department for more information.

As a licensed General Building Contractor in the state of California, I run into some poorly built stairways, every once in a while. They can and often do create safety hazards for anyone using them.

If you're building a new stairway, you will probably need a building permit, in most cities. However, if you're replacing a stairway, you might not need one, but it wouldn't be a bad idea to check with your local Building and Safety Department, before starting the project.

Let me get right to the point, the last thing you need is to build a stairway that creates a safety problem or needs to be removed and rebuilt, because it wasn't built correctly. Your local building and safety department could provide you with all sorts of great advice and make it well worth your time.

Notes

Installing Stair Treads

In the next step you will need to measure the distance in between the stair stringers, so that you can cut the stair treads. Remember, in our example we're using 2 x 12 stair treads, but you might need to use larger ones. (3 x 12, 4 x 12, ect.)

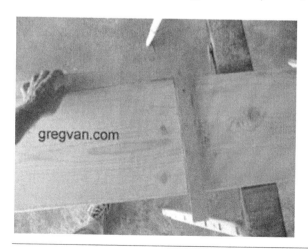

You should always use a framing square to mark your stair treads.

Be careful, marking your stair treads. If they're marked and cut incorrectly, you will have problems installing them.

It doesn't get any easier than this. After you've cut your stair treads, place them into their proper positions.

Notes
_____
_____
_____
_____
_____

_____

Then line the front of the stair treads up with the marks you made on the stringer. You can install one stair tread at a time or place all of them in their proper positions, before attaching them to the stair tread brackets.

Just don't stand on the stair treads, until there safely attached.

Notes

Attaching Stair Treads

The next step will be to attach your stair treads using the correct lag screws. Remember, some manufacturers will want you to use screws they recommend. Work your way up the stairway, installing the correct screws into each bracket.

These particular stair tread brackets call out for three screws to attach the stair bracket to the stringer and two screws to attach the stair tread to the bracket. If one or both of these screws are missing, for whatever reason, you could have a potential safety hazard.

If you're having a problem, installing the screws, you can always place the stair treads into the proper position, mark the holes and then pre-drill the treads. If any of the lag screws break and cannot be removed from the stair tread, then simply remove the stair tread and turn it around. Hopefully, once you turn the stair tread around, the broken lag screws that are stuck inside of the stair tread you're planning on using, won't line up with the same bracket holes.

Upper Deck Nosing

Even though I already mentioned it, you need to make sure the stair nosing on your landing, deck or floor is properly positioned.

Notes

Finished Stairway

If you completed your stairway, you can now consider yourself a stair builder.

Again, congratulations on your success.

We would love to see pictures of your completed project and get a little feedback about our stair building book. Our email address can be found on this website.

http://www.homebuildingandrepairs.com

Stair Building Glossary

Visit our online glossary for more information or if there's a word we might have missed at

http://stairs4u.com/glossary.htm

Anchor Bolts – Bolts that are embedded in the building foundation and used to attach building materials, like lumber to the foundation. Most anchor bolts are L-shaped and between 1/2 and 3/4" in diameter.

Balusters - These are vertical handrail components used to protect people from falling through the handrail or guard rail system. Balusters can be made from wood, metal and plastic.

Building Codes - A system of rules and regulations to create safer buildings. Most cities or counties throughout the United States regulate building codes through their building departments using building inspectors and other city officials.

Building Foundation - This is usually what a house or building rests upon. Foundations are usually built using concrete, block, cement or brick.

Building Hardware - Any nails, screws, nuts, bolts, hangers, connectors or other metal brackets used to connect one building component to another.

Cap - A piece of finish building material that usually runs at an angle, vertical or horizontal. Most solid drywall guardrails will use a 1 x 6 cap to finish the top of a wall off.

Cedar – Is a recommended wood used for exterior building projects like decks and stairs. Cedar can also be used on the interior of a house. Also see Redwood.

Ceiling - The overhead surface of a room, hallway and stairway. It's important to have enough distance between the ceiling and your head, for anyone who'll be walking up and down the stairs.

Construction Standard Lumber - Wood that is used primarily for framing and building homes. On the west coast of the United States they use Douglas Fir and on the East Coast they often use Southern Pine.

Diameter - The width of a circle, using a straight line that runs directly through the center.

Douglas Fir – A tall straight growing tree that provides plenty of lumber used for building homes throughout most western areas of the United States. Most construction standard lumber used in Southern California is Douglas Fir.

Drywall Spacer - Usually a 2 x 4 or 1 x 4 that attaches to one side of the stair stringer to allow a gap between the wall and the stringer to allow easy drywall installation around the stair stringer's.

First Floor - This would be the lower level of the building. The first floor would be a lower living area of a two-story house. The bottom of the stairway usually attaches to the first floor.

Framing Anchors - These can be made from lead, plastic or metal and are used to attach one type of building material to another. A good example of this would be using mechanical anchoring systems to attach treated lumber to a concrete building foundation.

Gripable Handrailing - This is the part of the handrailing system that's used to hold on to, while you're walking up and down a stairway. Most gripable handrails are located between 34 and 38 inches from the top of the stair tread nosing and run the entire length of the stairway.

Guard Railing - Consists of balusters, handrailing and posts used to create a protective barrier to prevent people from falling from heights exceeding 30 inches, above the lower level.

Head Room - This is the vertical distance between the stair tread nosing and the ceiling above. This building code prevents taller people from hitting their heads on the ceiling, as they walk up and down the stairway.

Headout - Is located in the upper level floor joist system, where the stairway attaches to the upper floor. The stair stringers usually attach to the headout or a ledger, nailed directly underneath the headout.

Jack Studs - These are vertical wall studs that form the walls underneath a stair stringer. The bottom cut of the jack studs are usually square, while the top cut is angled.

Joist - These are the horizontal structural support members of the floor framing system that are used to support the floor. Floor joist are usually 2 x 6, 2 x 8, 2 x 10, 2 x 12 or 2 x 14 and also come in the form of engineered lumber, like truss joist.

Joist Hanger - These are the metal connectors used to connect the floor joist to another structural component of the floor joist system, like a beam. You can also use joist hangers to connect the stair stringers to a ledger or headout.

Lag Screws - Are screw type bolts with pointed tips that are used to attach building components together. Stair builders often use lag screws to attach 3 x 12 treads to 4 x 12 stringers, using special brackets.

Landing - A floor located somewhere in a stairway. A stairway could go up five steps and then turn 90° at a stair landing and then continue up to the second floor. A stair landing could also be located at the bottom or top of the stairway.

Layout - Marking, measuring and planning out the design of walls, floors, roofs or stairway components. Laying out a stair stringer is often referred to as the process for planning and creating the stair stringer, before cutting it.

Ledger - This is usually a horizontal building component that's used to attach two sections of the building together. A good example of this would be using a ledger that attaches to a wall, providing something solid for your stair stringers to attach to.

Nails – Pin shaped pieces of metal that comes in different sizes and can be used to attach wood together. Most stairway framers use 16-d and 8-d nails for stair building.

Newel Post - Usually decorative wood posts used for supporting the handrailing system, located at the top, middle or bottom of a stairway. Newel posts often provide structural support for stairway handrails.

Nosing - This is the front of the stair tread or step. Some stair steps have a nosing that protrudes 1 inch away from the riser, while others don't protrude out at all, yet angle back in.

O.S.B. - Is often referred to as oriented strand board. This is a man-made product that requires gluing strands of lumber together, to produce an alternative building product to plywood.

Pattern - An original design created for the purpose of making copies. You can use patterns to make copies of stair stringers, treads and risers to speed up the stair assembly process.

Plywood - Is a man-made product made by gluing thin layers of lumber together with each layer changing direction at 90° angles, increasing the strength of the product as additional layers are glued to it. Plywood is often used for floors, structural walls, stair treads and risers.

Posts - Structural components that provide support for a beam or strength for handrails. Guard rails usually have posts built into them, to provide additional strength.

Radius - Is half the diameter of a circle. The radius is also the measurement from the center to the outside of a circle.

Railing - This is normally referred to as the top of the handrail or guard rail system. If you were walking next to a guard rail, you could run your fingers along the top of the railing.

Redwood - Is a preferred wood used for building outdoor decks and stairways. Redwood comes in different grades and some of these grades are extremely expensive. I've heard other people say that termites won't eat Redwood, but that's not true.

Risers - These are the vertical sections of the stairway in between the treads or steps. Risers can be made from a variety of materials, but the most common are made from construction standard, hardwoods, soft woods, plywood's and oriented strand board.

Screw - These are pieces of pin shaped metal with a threaded body that are often used to assemble home building products. Screws are often used instead of nails to provide additional holding power. Don't replace screws with nails, unless you have permission from the building designer, architect or engineer. Screws usually don't have the same sheer value a nail does.

Second Floor - This is usually the second level of a building. The top of the first stairway will attach to the second floor, while the bottom of the stairway attaches to the first level or floor. If you have a three level house, the bottom of the second set of stairs will attach to the second floor, while the top of the second stairway attaches to the third floor.

Sheathing - This is the material you will stand on, while walking around on a wood framed floor system. Sheathing is often referred to as underlayment. Most home builders use plywood or oriented strand board for sheathing.

Skirt Board - There are two types of skirt boards. The first type goes on the outside of the stair stringer and the second type separates your finished wall from the stairway. Both types of skirt boards are finished building products and are normally stained or painted.

Span - This would be the distance in between structural supports. The span will determine the thickness of stair treads, risers and floor sheathing.

Stairway - A passageway used to gain access from one level of the building to another. To get from the first floor of most houses to the second floor, a stairway is provided by the home builder for easy access.

Stairwell - This is an interior shape that's usually cut into a buildings floor framing system, to provide access to the second floor. The framing carpenters who are responsible for building the second floor are also responsible for building the stairwell.

Stringer - The main structural support for almost any stairway. Stair stringers can be made out of wood or metal and usually support each tread and riser (step).

Stringer Layout – This is the process for marking out each individual stair tread and riser, on the stair stringer. This usually requires a framing square, along with the height of each riser and the width of each stair tread.

Total Rise - This is the vertical distance from the top of the bottom floor to the top of the top floor. If you had a 100 inch measurement from the top of the building foundation (lower level), to the top of the second floor sheathing, then the total rise would be 100 inches.

Total Run - This is the horizontal distance from the front of the first stair step, to the back of the last stair step. If you went horizontally from the stair headout, to the first floor and measured the distance to the first step, this would be the total run.

Tread Brackets - These are brackets used to attach stair treads to stair stringers. Most of these tread brackets can be purchased from your local home improvement center or lumber yard.

Tread Overhang - This is the distance each stair tread overhangs from the riser. Most building codes won't let you have a tread overhang longer than 1 - 1/4 inches.

Tread Under Cut - This is the distance the stair tread protrudes into the stair riser. Instead of an overhang at the front of the stair step, the riser will be angled back, at the bottom, providing more room for the stair step or tread.

Tread – Another word for the stair step, often used by building professionals. Stair treads can be made from a variety of materials, including concrete, wood and metal.

Treated Lumber - Is wood that has been pressure-treated with chemicals to reduce the chances of wood decay. Treated lumber is often used at the bottom of wood framed walls and stair stringers, to reduce the chances of wood rot or decay.

Visit our online glossary for more information or if there's a word we might have missed at:

http://stairs4u.com/glossary.htm

Rise and Run Chart for 10 Inch Treads

Total Rise = Total height of stairway, from the top of the bottom floor, to the top of the top floor.

Risers = Amount of risers in stairway.

Riser Height = Individual riser height or total height between steps.

Steps = Amount of steps in stairway.

Total Run = Length of stairway in inches, using a 10 inch step or stair tread.

Angle = This is the angle of the stairway or incline.

Total Rise	Risers	Riser Height	Steps	Total Run	Angle
10	2	5.00	1	10	26.57
10.25	2	5.13	1	10	27.14
10.5	2	5.25	1	10	27.70
10.75	2	5.38	1	10	28.26
10.75	2	5.38	1	10	28.26
11.25	2	5.63	1	10	29.36
11.5	2	5.75	1	10	29.90
11.75	2	5.88	1	10	30.43
12	2	6.00	1	10	30.96
12.25	2	6.13	1	10	31.49
12.5	2	6.25	1	10	32.01
12.75	2	6.38	1	10	32.52
13	2	6.50	1	10	33.02
13.25	2	6.63	1	10	33.52
13.5	2	6.75	1	10	34.02
13.75	2	6.88	1	10	34.51
14	2	7.00	1	10	34.99
14.25	2	7.13	1	10	35.47
14.5	2	7.25	1	10	35.94
14.75	2	7.38	1	10	36.41

Total Rise	Risers	Riser Height	Steps	Total Run	Angle
15	2	7.50	1	10	36.87
15.25	2	7.63	1	10	37.33
15.5	2	7.75	1	10	37.78
15.75	3	5.25	2	20	27.70
16	3	5.33	2	20	28.07
16.25	3	5.42	2	20	28.44
16.5	3	5.50	2	20	28.81
16.75	3	5.58	2	20	29.18
17	3	5.67	2	20	29.54
17.25	3	5.75	2	20	29.90
17.5	3	5.83	2	20	30.26
17.75	3	5.92	2	20	30.61
18	3	6.00	2	20	30.96
18.25	3	6.08	2	20	31.31
18.5	3	6.17	2	20	31.66
18.75	3	6.25	2	20	32.01
19	3	6.33	2	20	32.35
19.25	3	6.42	2	20	32.69
19.5	3	6.50	2	20	33.02
19.75	3	6.58	2	20	33.36
20	3	6.67	2	20	33.69
20.25	3	6.75	2	20	34.02
20.5	3	6.83	2	20	34.35
20.75	3	6.92	2	20	34.67
21	3	7.00	2	20	34.99
21.25	3	7.08	2	20	35.31
21.5	3	7.17	2	20	35.63
21.75	3	7.25	2	20	35.94
22	3	7.33	2	20	36.25
22.25	3	7.42	2	20	36.56
22.5	3	7.50	2	20	36.87
22.75	3	7.58	2	20	37.17
23	3	7.67	2	20	37.48
23.25	3	7.75	2	20	37.78
23.5	4	5.88	3	30	30.43
23.75	4	5.94	3	30	30.70
24	4	6.00	3	30	30.96
24.25	4	6.06	3	30	31.23
24.5	4	6.13	3	30	31.49
24.75	4	6.19	3	30	31.75
25	4	6.25	3	30	32.01

Total Rise	Risers	Riser Height	Steps	Total Run	Angle
25.25	4	6.31	3	30	32.26
25.5	4	6.38	3	30	32.52
25.75	4	6.44	3	30	32.77
26	4	6.50	3	30	33.02
26.25	4	6.56	3	30	33.27
26.5	4	6.63	3	30	33.52
26.75	4	6.69	3	30	33.77
27	4	6.75	3	30	34.02
27.25	4	6.81	3	30	34.26
27.5	4	6.88	3	30	34.51
27.75	4	6.94	3	30	34.75
28	4	7.00	3	30	34.99
28.25	4	7.06	3	30	35.23
28.5	4	7.13	3	30	35.47
28.75	4	7.19	3	30	35.71
29	4	7.25	3	30	35.94
29.25	4	7.31	3	30	36.18
29.5	4	7.38	3	30	36.41
29.75	4	7.44	3	30	36.64
30	4	7.50	3	30	36.87
30.25	4	7.56	3	30	37.10
30.5	4	7.63	3	30	37.33
30.75	4	7.69	3	30	37.55
31	4	7.75	3	30	37.78
31.25	5	6.25	4	40	32.01
31.5	5	6.30	4	40	32.21
31.75	5	6.35	4	40	32.42
32	5	6.40	4	40	32.62
32.25	5	6.45	4	40	32.82
32.5	5	6.50	4	40	33.02
32.75	5	6.55	4	40	33.22
33	5	6.60	4	40	33.42
33.25	5	6.65	4	40	33.62
33.5	5	6.70	4	40	33.82
33.75	5	6.75	4	40	34.02
34	5	6.80	4	40	34.22
34.25	5	6.85	4	40	34.41
34.5	5	6.90	4	40	34.61
34.75	5	6.95	4	40	34.80
35	5	7.00	4	40	34.99
35.25	5	7.05	4	40	35.18

Total Rise	Risers	Riser Height	Steps	Total Run	Angle
35.5	5	7.10	4	40	35.37
35.75	5	7.15	4	40	35.56
36	5	7.20	4	40	35.75
36.25	5	7.25	4	40	35.94
36.5	5	7.30	4	40	36.13
36.75	5	7.35	4	40	36.32
37	5	7.40	4	40	36.50
37.25	5	7.45	4	40	36.69
37.5	5	7.50	4	40	36.87
37.75	5	7.55	4	40	37.05
38	5	7.60	4	40	37.23
38.25	5	7.65	4	40	37.42
38.5	5	7.70	4	40	37.60
38.75	5	7.75	4	40	37.78
39	6	6.50	5	50	33.02
39.25	6	6.54	5	50	33.19
39.5	6	6.58	5	50	33.36
39.75	6	6.63	5	50	33.52
40	6	6.67	5	50	33.69
40.25	6	6.71	5	50	33.86
40.5	6	6.75	5	50	34.02
40.75	6	6.79	5	50	34.18
41	6	6.83	5	50	34.35
41.25	6	6.88	5	50	34.51
41.5	6	6.92	5	50	34.67
41.75	6	6.96	5	50	34.83
42	6	7.00	5	50	34.99
42.25	6	7.04	5	50	35.15
42.5	6	7.08	5	50	35.31
42.75	6	7.13	5	50	35.47
43	6	7.17	5	50	35.63
43.25	6	7.21	5	50	35.79
43.5	6	7.25	5	50	35.94
43.75	6	7.29	5	50	36.10
44	6	7.33	5	50	36.25
44.25	6	7.38	5	50	36.41
44.5	6	7.42	5	50	36.56
44.75	6	7.46	5	50	36.72
45	6	7.50	5	50	36.87
45.25	6	7.54	5	50	37.02
45.5	6	7.58	5	50	37.17

Total Rise	Risers	Riser Height	Steps	Total Run	Angle
45.75	6	7.63	5	50	37.33
46	6	7.67	5	50	37.48
46.25	6	7.71	5	50	37.63
46.5	6	7.75	5	50	37.78
46.75	7	6.68	6	60	33.74
47	7	6.71	6	60	33.88
47.25	7	6.75	6	60	34.02
47.5	7	6.79	6	60	34.16
47.75	7	6.82	6	60	34.30
48	7	6.86	6	60	34.44
48.25	7	6.89	6	60	34.58
48.5	7	6.93	6	60	34.72
48.75	7	6.96	6	60	34.85
49	7	7.00	6	60	34.99
49.25	7	7.04	6	60	35.13
49.5	7	7.07	6	60	35.27
49.75	7	7.11	6	60	35.40
50	7	7.14	6	60	35.54
5.25	7	0.75	6	60	4.29
50.5	7	7.21	6	60	35.81
50.75	7	7.25	6	60	35.94
51	7	7.29	6	60	36.08
51.25	7	7.32	6	60	36.21
51.5	7	7.36	6	60	36.34
51.75	7	7.39	6	60	36.47
52	7	7.43	6	60	36.61
52.25	7	7.46	6	60	36.74
52.5	7	7.50	6	60	36.87
52.75	7	7.54	6	60	37.00
53	7	7.57	6	60	37.13
53.25	7	7.61	6	60	37.26
53.5	7	7.64	6	60	37.39
53.75	7	7.68	6	60	37.52
54	7	7.71	6	60	37.65
54.25	7	7.75	6	60	37.78
54.5	8	6.81	7	70	34.26
54.75	8	6.84	7	70	34.39
55	8	6.88	7	70	34.51
55.25	8	6.91	7	70	34.63
55.5	8	6.94	7	70	34.75
55.75	8	6.97	7	70	34.87

Total Rise	Risers	Riser Height	Steps	Total Run	Angle
56	8	7.00	7	70	34.99
56.25	8	7.03	7	70	35.11
56.5	8	7.06	7	70	35.23
56.75	8	7.09	7	70	35.35
57	8	7.13	7	70	35.47
57.25	8	7.16	7	70	35.59
57.5	8	7.19	7	70	35.71
57.75	8	7.22	7	70	35.82
58	8	7.25	7	70	35.94
58.25	8	7.28	7	70	36.06
58.5	8	7.31	7	70	36.18
58.75	8	7.34	7	70	36.29
59	8	7.38	7	70	36.41
59.25	8	7.41	7	70	36.52
59.5	8	7.44	7	70	36.64
59.75	8	7.47	7	70	36.76
60	8	7.50	7	70	36.87
60.25	8	7.53	7	70	36.98
60.5	8	7.56	7	70	37.10
60.75	8	7.59	7	70	37.21
61	8	7.63	7	70	37.33
61.25	8	7.66	7	70	37.44
61.5	8	7.69	7	70	37.55
61.75	8	7.72	7	70	37.66
62	8	7.75	7	70	37.78
62.25	9	6.92	8	80	34.67
62.5	9	6.94	8	80	34.78
62.75	9	6.97	8	80	34.89
63	9	7.00	8	80	34.99
63.25	9	7.03	8	80	35.10
63.5	9	7.06	8	80	35.21
63.75	9	7.08	8	80	35.31
64	9	7.11	8	80	35.42
64.25	9	7.14	8	80	35.52
64.5	9	7.17	8	80	35.63
64.75	9	7.19	8	80	35.73
65	9	7.22	8	80	35.84
65.25	9	7.25	8	80	35.94
65.5	9	7.28	8	80	36.05
65.75	9	7.31	8	80	36.15
66	9	7.33	8	80	36.25

Total Rise	Risers	Riser Height	Steps	Total Run	Angle
66.25	9	7.36	8	80	36.36
66.5	9	7.39	8	80	36.46
66.75	9	7.42	8	80	36.56
67	9	7.44	8	80	36.67
67.25	9	7.47	8	80	36.77
67.5	9	7.50	8	80	36.87
67.75	9	7.53	8	80	36.97
68	9	7.56	8	80	37.07
68.25	9	7.58	8	80	37.17
68.5	9	7.61	8	80	37.28
68.75	9	7.64	8	80	37.38
69	10	6.90	9	90	34.61
69.25	10	6.93	9	90	34.70
69.5	10	6.95	9	90	34.80
69.75	10	6.98	9	90	34.90
70	10	7.00	9	90	34.99
70.25	10	7.03	9	90	35.09
70.5	10	7.05	9	90	35.18
70.75	10	7.08	9	90	35.28
71	10	7.10	9	90	35.37
71.25	10	7.13	9	90	35.47
71.5	10	7.15	9	90	35.56
71.75	10	7.18	9	90	35.66
72	10	7.20	9	90	35.75
72.25	10	7.23	9	90	35.85
72.5	10	7.25	9	90	35.94
72.75	10	7.28	9	90	36.04
73	10	7.30	9	90	36.13
73.25	10	7.33	9	90	36.22
73.5	10	7.35	9	90	36.32
73.75	10	7.38	9	90	36.41
74	10	7.40	9	90	36.50
74.25	10	7.43	9	90	36.59
74.5	10	7.45	9	90	36.69
74.75	10	7.48	9	90	36.78
75	10	7.50	9	90	36.87
75.25	10	7.53	9	90	36.96
75.5	10	7.55	9	90	37.05
75.75	10	7.58	9	90	37.14
76	10	7.60	9	90	37.23
76.25	10	7.63	9	90	37.33

Total Rise	Risers	Riser Height	Steps	Total Run	Angle
76.5	11	6.95	10	100	34.82
76.75	11	6.98	10	100	34.90
77	11	7.00	10	100	34.99
77.25	11	7.02	10	100	35.08
77.5	11	7.05	10	100	35.17
77.75	11	7.07	10	100	35.25
78	11	7.09	10	100	35.34
78.25	11	7.11	10	100	35.43
78.5	11	7.14	10	100	35.51
78.75	11	7.16	10	100	35.60
79	11	7.18	10	100	35.69
79.25	11	7.20	10	100	35.77
79.5	11	7.23	10	100	35.86
79.75	11	7.25	10	100	35.94
80	11	7.27	10	100	36.03
80.25	11	7.30	10	100	36.11
80.5	11	7.32	10	100	36.20
80.75	11	7.34	10	100	36.28
81	11	7.36	10	100	36.37
81.25	11	7.39	10	100	36.45
81.5	11	7.41	10	100	36.54
81.75	11	7.43	10	100	36.62
82	11	7.45	10	100	36.70
82.25	11	7.48	10	100	36.79
82.5	11	7.50	10	100	36.87
82.75	11	7.52	10	100	36.95
83	11	7.55	10	100	37.04
83.25	11	7.57	10	100	37.12
83.5	11	7.59	10	100	37.20
83.75	11	7.61	10	100	37.28
84	11	7.64	10	100	37.37
84.25	12	7.02	11	110	35.07
84.5	12	7.04	11	110	35.15
84.75	12	7.06	11	110	35.23
85	12	7.08	11	110	35.31
85.25	12	7.10	11	110	35.39
85.5	12	7.13	11	110	35.47
85.75	12	7.15	11	110	35.55
86	12	7.17	11	110	35.63
86.25	12	7.19	11	110	35.71
86.5	12	7.21	11	110	35.79

Total Rise	Risers	Riser Height	Steps	Total Run	Angle
86.75	12	7.23	11	110	35.86
87	12	7.25	11	110	35.94
87.25	12	7.27	11	110	36.02
87.5	12	7.29	11	110	36.10
87.75	12	7.31	11	110	36.18
88	12	7.33	11	110	36.25
88.25	12	7.35	11	110	36.33
88.5	12	7.38	11	110	36.41
88.75	12	7.40	11	110	36.49
89	12	7.42	11	110	36.56
89.25	12	7.44	11	110	36.64
89.5	12	7.46	11	110	36.72
89.75	12	7.48	11	110	36.79
90	12	7.50	11	110	36.87
90.25	12	7.52	11	110	36.95
90.5	12	7.54	11	110	37.02
90.75	12	7.56	11	110	37.10
91	12	7.58	11	110	37.17
91.25	13	7.02	12	120	35.07
91.5	13	7.04	12	120	35.14
91.75	13	7.06	12	120	35.21
92	13	7.08	12	120	35.29
92.25	13	7.10	12	120	35.36
92.5	13	7.12	12	120	35.43
92.75	13	7.13	12	120	35.51
93	13	7.15	12	120	35.58
93.25	13	7.17	12	120	35.65
93.5	13	7.19	12	120	35.72
93.75	13	7.21	12	120	35.80
94	13	7.23	12	120	35.87
94.25	13	7.25	12	120	35.94
94.5	13	7.27	12	120	36.01
94.75	13	7.29	12	120	36.09
95	13	7.31	12	120	36.16
95.25	13	7.33	12	120	36.23
95.5	13	7.35	12	120	36.30
95.75	13	7.37	12	120	36.37
96	13	7.38	12	120	36.44
96.25	13	7.40	12	120	36.52
96.5	13	7.42	12	120	36.59
96.75	13	7.44	12	120	36.66

Total Rise	Risers	Riser Height	Steps	Total Run	Angle
97	13	7.46	12	120	36.73
97.25	13	7.48	12	120	36.80
97.5	13	7.50	12	120	36.87
97.75	13	7.52	12	120	36.94
98	14	7.00	13	130	34.99
98.25	14	7.02	13	130	35.06
98.5	14	7.04	13	130	35.13
98.75	14	7.05	13	130	35.20
99	14	7.07	13	130	35.27
99.25	14	7.09	13	130	35.33
99.5	14	7.11	13	130	35.40
99.75	14	7.13	13	130	35.47
100	14	7.14	13	130	35.54
100.25	14	7.16	13	130	35.61
100.5	14	7.18	13	130	35.67
100.75	14	7.20	13	130	35.74
101	14	7.21	13	130	35.81
101.25	14	7.23	13	130	35.87
101.5	14	7.25	13	130	35.94
101.75	14	7.27	13	130	36.01
102	14	7.29	13	130	36.08
102.25	14	7.30	13	130	36.14
102.5	14	7.32	13	130	36.21
102.75	14	7.34	13	130	36.28
103	14	7.36	13	130	36.34
103.25	14	7.38	13	130	36.41
103.5	14	7.39	13	130	36.47
103.75	14	7.41	13	130	36.54
104	14	7.43	13	130	36.61
104.25	14	7.45	13	130	36.67
104.5	14	7.46	13	130	36.74
104.75	14	7.48	13	130	36.80
105	14	7.50	13	130	36.87
105.25	15	7.02	14	140	35.06
105.5	15	7.03	14	140	35.12
105.75	15	7.05	14	140	35.18
106	15	7.07	14	140	35.25
106.25	15	7.08	14	140	35.31
106.5	15	7.10	14	140	35.37
106.75	15	7.12	14	140	35.44
107	15	7.13	14	140	35.50

Total Rise	Risers	Riser Height	Steps	Total Run	Angle
107.25	15	7.15	14	140	35.56
107.5	15	7.17	14	140	35.63
107.75	15	7.18	14	140	35.69
108	15	7.20	14	140	35.75
108.25	15	7.22	14	140	35.82
108.5	15	7.23	14	140	35.88
108.75	15	7.25	14	140	35.94
109	15	7.27	14	140	36.00
109.25	15	7.28	14	140	36.07
109.5	15	7.30	14	140	36.13
109.75	15	7.32	14	140	36.19
110	15	7.33	14	140	36.25
110.25	15	7.35	14	140	36.32
110.5	15	7.37	14	140	36.38
110.75	15	7.38	14	140	36.44
111	15	7.40	14	140	36.50
111.25	15	7.42	14	140	36.56
111.5	15	7.43	14	140	36.62
111.75	15	7.45	14	140	36.69
112	15	7.47	14	140	36.75
112.25	15	7.48	14	140	36.81
112.5	15	7.50	14	140	36.87
112.75	16	7.05	15	150	35.17
113	16	7.06	15	150	35.23
113.25	16	7.08	15	150	35.29
113.5	16	7.09	15	150	35.35
113.75	16	7.11	15	150	35.41
114	16	7.13	15	150	35.47
114.25	16	7.14	15	150	35.53
114.5	16	7.16	15	150	35.59
114.75	16	7.17	15	150	35.65
115	16	7.19	15	150	35.71
115.25	16	7.20	15	150	35.77
115.5	16	7.22	15	150	35.82
115.75	16	7.23	15	150	35.88
116	16	7.25	15	150	35.94
116.25	16	7.27	15	150	36.00
116.5	16	7.28	15	150	36.06
116.75	16	7.30	15	150	36.12
117	16	7.31	15	150	36.18
117.25	16	7.33	15	150	36.23

Total Rise	Risers	Riser Height	Steps	Total Run	Angle
117.5	16	7.34	15	150	36.29
117.75	16	7.36	15	150	36.35
118	16	7.38	15	150	36.41
118.25	16	7.39	15	150	36.47
118.5	16	7.41	15	150	36.52
118.75	16	7.42	15	150	36.58
119	16	7.44	15	150	36.64
119.25	16	7.45	15	150	36.70
119.5	16	7.47	15	150	36.76
119.75	16	7.48	15	150	36.81
120	16	7.50	15	150	36.87
120.25	17	7.07	16	160	35.27
120.5	17	7.09	16	160	35.33
120.75	17	7.10	16	160	35.39
121	17	7.12	16	160	35.44
121.25	17	7.13	16	160	35.50
121.5	17	7.15	16	160	35.55
121.75	17	7.16	16	160	35.61
122	17	7.18	16	160	35.67
122.25	17	7.19	16	160	35.72
122.5	17	7.21	16	160	35.78
122.75	17	7.22	16	160	35.83
123	17	7.24	16	160	35.89
123.25	17	7.25	16	160	35.94
123.5	17	7.26	16	160	36.00
123.75	17	7.28	16	160	36.05
124	17	7.29	16	160	36.11
124.25	17	7.31	16	160	36.16
124.5	17	7.32	16	160	36.22
124.75	17	7.34	16	160	36.27
125	17	7.35	16	160	36.33
125.25	17	7.37	16	160	36.38
125.5	17	7.38	16	160	36.44
125.75	17	7.40	16	160	36.49
126	17	7.41	16	160	36.54
126.25	17	7.43	16	160	36.60
126.5	17	7.44	16	160	36.65
126.75	17	7.46	16	160	36.71
127	17	7.47	16	160	36.76
127.25	17	7.49	16	160	36.82
127.5	17	7.50	16	160	36.87

Total Rise	Risers	Riser Height	Steps	Total Run	Angle
127.75	18	7.10	17	170	35.36
128	18	7.11	17	170	35.42
128.25	18	7.13	17	170	35.47
128.5	18	7.14	17	170	35.52
128.75	18	7.15	17	170	35.58
129	18	7.17	17	170	35.63
129.25	18	7.18	17	170	35.68
129.5	18	7.19	17	170	35.73
129.75	18	7.21	17	170	35.79
130	18	7.22	17	170	35.84
130.25	18	7.24	17	170	35.89
130.5	18	7.25	17	170	35.94
130.75	18	7.26	17	170	35.99
131	18	7.28	17	170	36.05
131.25	18	7.29	17	170	36.10
131.5	18	7.31	17	170	36.15
131.75	18	7.32	17	170	36.20
132	18	7.33	17	170	36.25
132.25	18	7.35	17	170	36.31
132.5	18	7.36	17	170	36.36
132.75	18	7.38	17	170	36.41
133	18	7.39	17	170	36.46
133.25	18	7.40	17	170	36.51
133.5	18	7.42	17	170	36.56
133.75	18	7.43	17	170	36.61
134	18	7.44	17	170	36.67
134.25	18	7.46	17	170	36.72
134.5	18	7.47	17	170	36.77
134.75	18	7.49	17	170	36.82
135	18	7.50	17	170	36.87
135.25	19	7.12	18	180	35.44
135.5	19	7.13	18	180	35.49
135.75	19	7.14	18	180	35.54
135	19	7.11	18	180	35.39
135.25	19	7.12	18	180	35.44
135.5	19	7.13	18	180	35.49
135.75	19	7.14	18	180	35.54
136	19	7.16	18	180	35.59
136.25	19	7.17	18	180	35.64
136.5	19	7.18	18	180	35.69
136.75	19	7.20	18	180	35.74

Total Rise	Risers	Riser Height	Steps	Total Run	Angle
137	19	7.21	18	180	35.79
137.25	19	7.22	18	180	35.84
137.5	19	7.24	18	180	35.89
137.75	19	7.25	18	180	35.94
138	19	7.26	18	180	35.99
138.25	19	7.28	18	180	36.04
138.5	19	7.29	18	180	36.09
138.75	19	7.30	18	180	36.14
139	19	7.32	18	180	36.19
139.25	19	7.33	18	180	36.24
139.5	19	7.34	18	180	36.29
139.75	19	7.36	18	180	36.34
140	19	7.37	18	180	36.38

Rise and Run Chart for 17 - ½ Inch Rule

Total Rise = Total height of stairway, from the top of the bottom floor to the top of the top floor.

Risers = Amount of risers in stairway.

Riser Height = Individual riser height or total height between steps.

Steps = Amount of steps in stairway.

Total Run = Length of stairway in inches, while subtracting height of riser from 17.5 inches.

Angle = This is the angle of the stairway or incline.

Total Rise	Risers	Riser Height	Steps	Tread	Total Run	Angle
10	2	5.00	1	12.50	12.50	21.80
10.25	2	5.13	1	12.38	12.38	22.50
10.5	2	5.25	1	12.25	12.25	23.20
10.75	2	5.38	1	12.13	12.13	23.91
10.75	2	5.38	1	12.13	12.13	23.91
11.25	2	5.63	1	11.88	11.88	25.35
11.5	2	5.75	1	11.75	11.75	26.08
11.75	2	5.88	1	11.63	11.63	26.81
12	2	6.00	1	11.50	11.50	27.55
12.25	2	6.13	1	11.38	11.38	28.30
12.5	2	6.25	1	11.25	11.25	29.05
12.75	2	6.38	1	11.13	11.13	29.81
13	2	6.50	1	11.00	11.00	30.58
13.25	2	6.63	1	10.88	10.88	31.35
13.5	2	6.75	1	10.75	10.75	32.12
13.75	2	6.88	1	10.63	10.63	32.91
14	2	7.00	1	10.50	10.50	33.69
14.25	2	7.13	1	10.38	10.38	34.48

Total Rise	Risers	Riser Height	Steps	Tread	Total Run	Angle
14.5	2	7.25	1	10.25	10.25	35.27
14.75	2	7.38	1	10.13	10.13	36.07
15	2	7.50	1	10.00	10.00	36.87
15.25	2	7.63	1	9.88	9.88	37.67
15.5	2	7.75	1	9.75	9.75	38.48
15.75	3	5.25	2	12.25	24.50	23.20
16	3	5.33	2	12.17	24.33	23.67
16.25	3	5.42	2	12.08	24.17	24.15
16.5	3	5.50	2	12.00	24.00	24.62
16.75	3	5.58	2	11.92	23.83	25.10
17	3	5.67	2	11.83	23.67	25.59
17.25	3	5.75	2	11.75	23.50	26.08
17.5	3	5.83	2	11.67	23.33	26.57
17.75	3	5.92	2	11.58	23.17	27.06
18	3	6.00	2	11.50	23.00	27.55
18.25	3	6.08	2	11.42	22.83	28.05
18.5	3	6.17	2	11.33	22.67	28.55
18.75	3	6.25	2	11.25	22.50	29.05
19	3	6.33	2	11.17	22.33	29.56
19.25	3	6.42	2	11.08	22.17	30.07
19.5	3	6.50	2	11.00	22.00	30.58
19.75	3	6.58	2	10.92	21.83	31.09
20	3	6.67	2	10.83	21.67	31.61
20.25	3	6.75	2	10.75	21.50	32.12
20.5	3	6.83	2	10.67	21.33	32.64
20.75	3	6.92	2	10.58	21.17	33.17
21	3	7.00	2	10.50	21.00	33.69
21.25	3	7.08	2	10.42	20.83	34.22
21.5	3	7.17	2	10.33	20.67	34.74
21.75	3	7.25	2	10.25	20.50	35.27
22	3	7.33	2	10.17	20.33	35.80
22.25	3	7.42	2	10.08	20.17	36.34
22.5	3	7.50	2	10.00	20.00	36.87
22.75	3	7.58	2	9.92	19.83	37.41
23	3	7.67	2	9.83	19.67	37.94
23.25	3	7.75	2	9.75	19.50	38.48
23.5	4	5.88	3	11.63	34.88	26.81
23.75	4	5.94	3	11.56	34.69	27.18
24	4	6.00	3	11.50	34.50	27.55

Total Rise	Risers	Riser Height	Steps	Tread	Total Run	Angle
24.25	4	6.06	3	11.44	34.31	27.93
24.5	4	6.13	3	11.38	34.13	28.30
24.75	4	6.19	3	11.31	33.94	28.68
25	4	6.25	3	11.25	33.75	29.05
25.25	4	6.31	3	11.19	33.56	29.43
25.5	4	6.38	3	11.13	33.38	29.81
25.75	4	6.44	3	11.06	33.19	30.20
26	4	6.50	3	11.00	33.00	30.58
26.25	4	6.56	3	10.94	32.81	30.96
26.5	4	6.63	3	10.88	32.63	31.35
26.75	4	6.69	3	10.81	32.44	31.74
27	4	6.75	3	10.75	32.25	32.12
27.25	4	6.81	3	10.69	32.06	32.51
27.5	4	6.88	3	10.63	31.88	32.91
27.75	4	6.94	3	10.56	31.69	33.30
28	4	7.00	3	10.50	31.50	33.69
28.25	4	7.06	3	10.44	31.31	34.08
28.5	4	7.13	3	10.38	31.13	34.48
28.75	4	7.19	3	10.31	30.94	34.88
29	4	7.25	3	10.25	30.75	35.27
29.25	4	7.31	3	10.19	30.56	35.67
29.5	4	7.38	3	10.13	30.38	36.07
29.75	4	7.44	3	10.06	30.19	36.47
30	4	7.50	3	10.00	30.00	36.87
30.25	4	7.56	3	9.94	29.81	37.27
30.5	4	7.63	3	9.88	29.63	37.67
30.75	4	7.69	3	9.81	29.44	38.08
31	4	7.75	3	9.75	29.25	38.48
31.25	5	6.25	4	11.25	45.00	29.05
31.5	5	6.30	4	11.20	44.80	29.36
31.75	5	6.35	4	11.15	44.60	29.66
32	5	6.40	4	11.10	44.40	29.97
32.25	5	6.45	4	11.05	44.20	30.27
32.5	5	6.50	4	11.00	44.00	30.58
32.75	5	6.55	4	10.95	43.80	30.89
33	5	6.60	4	10.90	43.60	31.20
33.25	5	6.65	4	10.85	43.40	31.50
33.5	5	6.70	4	10.80	43.20	31.81
33.75	5	6.75	4	10.75	43.00	32.12

Total Rise	Risers	Riser Height	Steps	Tread	Total Run	Angle
34	5	6.80	4	10.70	42.80	32.44
34.25	5	6.85	4	10.65	42.60	32.75
34.5	5	6.90	4	10.60	42.40	33.06
34.75	5	6.95	4	10.55	42.20	33.38
35	5	7.00	4	10.50	42.00	33.69
35.25	5	7.05	4	10.45	41.80	34.01
35.5	5	7.10	4	10.40	41.60	34.32
35.75	5	7.15	4	10.35	41.40	34.64
36	5	7.20	4	10.30	41.20	34.95
36.25	5	7.25	4	10.25	41.00	35.27
36.5	5	7.30	4	10.20	40.80	35.59
36.75	5	7.35	4	10.15	40.60	35.91
37	5	7.40	4	10.10	40.40	36.23
37.25	5	7.45	4	10.05	40.20	36.55
37.5	5	7.50	4	10.00	40.00	36.87
37.75	5	7.55	4	9.95	39.80	37.19
38	5	7.60	4	9.90	39.60	37.51
38.25	5	7.65	4	9.85	39.40	37.83
38.5	5	7.70	4	9.80	39.20	38.16
38.75	5	7.75	4	9.75	39.00	38.48
39	6	6.50	5	11.00	55.00	30.58
39.25	6	6.54	5	10.96	54.79	30.84
39.5	6	6.58	5	10.92	54.58	31.09
39.75	6	6.63	5	10.88	54.38	31.35
40	6	6.67	5	10.83	54.17	31.61
40.25	6	6.71	5	10.79	53.96	31.87
40.5	6	6.75	5	10.75	53.75	32.12
40.75	6	6.79	5	10.71	53.54	32.38
41	6	6.83	5	10.67	53.33	32.64
41.25	6	6.88	5	10.63	53.13	32.91
41.5	6	6.92	5	10.58	52.92	33.17
41.75	6	6.96	5	10.54	52.71	33.43
42	6	7.00	5	10.50	52.50	33.69
42.25	6	7.04	5	10.46	52.29	33.95
42.5	6	7.08	5	10.42	52.08	34.22
42.75	6	7.13	5	10.38	51.88	34.48
43	6	7.17	5	10.33	51.67	34.74
43.25	6	7.21	5	10.29	51.46	35.01
43.5	6	7.25	5	10.25	51.25	35.27

Total Rise	Risers	Riser Height	Steps	Tread	Total Run	Angle
43.75	6	7.29	5	10.21	51.04	35.54
44	6	7.33	5	10.17	50.83	35.80
44.25	6	7.38	5	10.13	50.63	36.07
44.5	6	7.42	5	10.08	50.42	36.34
44.75	6	7.46	5	10.04	50.21	36.60
45	6	7.50	5	10.00	50.00	36.87
45.25	6	7.54	5	9.96	49.79	37.14
45.5	6	7.58	5	9.92	49.58	37.41
45.75	6	7.63	5	9.88	49.38	37.67
46	6	7.67	5	9.83	49.17	37.94
46.25	6	7.71	5	9.79	48.96	38.21
46.5	6	7.75	5	9.75	48.75	38.48
46.75	7	6.68	6	10.82	64.93	31.68
47	7	6.71	6	10.79	64.71	31.90
47.25	7	6.75	6	10.75	64.50	32.12
47.5	7	6.79	6	10.71	64.29	32.35
47.75	7	6.82	6	10.68	64.07	32.57
48	7	6.86	6	10.64	63.86	32.79
48.25	7	6.89	6	10.61	63.64	33.02
48.5	7	6.93	6	10.57	63.43	33.24
48.75	7	6.96	6	10.54	63.21	33.47
49	7	7.00	6	10.50	63.00	33.69
49.25	7	7.04	6	10.46	62.79	33.92
49.5	7	7.07	6	10.43	62.57	34.14
49.75	7	7.11	6	10.39	62.36	34.37
50	7	7.14	6	10.36	62.14	34.59
5.25	7	0.75	6	16.75	100.50	2.56
50.5	7	7.21	6	10.29	61.71	35.05
50.75	7	7.25	6	10.25	61.50	35.27
51	7	7.29	6	10.21	61.29	35.50
51.25	7	7.32	6	10.18	61.07	35.73
51.5	7	7.36	6	10.14	60.86	35.96
51.75	7	7.39	6	10.11	60.64	36.18
52	7	7.43	6	10.07	60.43	36.41
52.25	7	7.46	6	10.04	60.21	36.64
52.5	7	7.50	6	10.00	60.00	36.87
52.75	7	7.54	6	9.96	59.79	37.10
53	7	7.57	6	9.93	59.57	37.33
53.25	7	7.61	6	9.89	59.36	37.56

Total Rise	Risers	Riser Height	Steps	Tread	Total Run	Angle
53.5	7	7.64	6	9.86	59.14	37.79
53.75	7	7.68	6	9.82	58.93	38.02
54	7	7.71	6	9.79	58.71	38.25
54.25	7	7.75	6	9.75	58.50	38.48
54.5	8	6.81	7	10.69	74.81	32.51
54.75	8	6.84	7	10.66	74.59	32.71
55	8	6.88	7	10.63	74.38	32.91
55.25	8	6.91	7	10.59	74.16	33.10
55.5	8	6.94	7	10.56	73.94	33.30
55.75	8	6.97	7	10.53	73.72	33.49
56	8	7.00	7	10.50	73.50	33.69
56.25	8	7.03	7	10.47	73.28	33.89
56.5	8	7.06	7	10.44	73.06	34.08
56.75	8	7.09	7	10.41	72.84	34.28
57	8	7.13	7	10.38	72.63	34.48
57.25	8	7.16	7	10.34	72.41	34.68
57.5	8	7.19	7	10.31	72.19	34.88
57.75	8	7.22	7	10.28	71.97	35.07
58	8	7.25	7	10.25	71.75	35.27
58.25	8	7.28	7	10.22	71.53	35.47
58.5	8	7.31	7	10.19	71.31	35.67
58.75	8	7.34	7	10.16	71.09	35.87
59	8	7.38	7	10.13	70.88	36.07
59.25	8	7.41	7	10.09	70.66	36.27
59.5	8	7.44	7	10.06	70.44	36.47
59.75	8	7.47	7	10.03	70.22	36.67
60	8	7.50	7	10.00	70.00	36.87
60.25	8	7.53	7	9.97	69.78	37.07
60.5	8	7.56	7	9.94	69.56	37.27
60.75	8	7.59	7	9.91	69.34	37.47
61	8	7.63	7	9.88	69.13	37.67
61.25	8	7.66	7	9.84	68.91	37.87
61.5	8	7.69	7	9.81	68.69	38.08
61.75	8	7.72	7	9.78	68.47	38.28
62	8	7.75	7	9.75	68.25	38.48
62.25	9	6.92	8	10.58	84.67	33.17
62.5	9	6.94	8	10.56	84.44	33.34
62.75	9	6.97	8	10.53	84.22	33.52
63	9	7.00	8	10.50	84.00	33.69

Total Rise	Risers	Riser Height	Steps	Tread	Total Run	Angle
63.25	9	7.03	8	10.47	83.78	33.87
63.5	9	7.06	8	10.44	83.56	34.04
63.75	9	7.08	8	10.42	83.33	34.22
64	9	7.11	8	10.39	83.11	34.39
64.25	9	7.14	8	10.36	82.89	34.57
64.5	9	7.17	8	10.33	82.67	34.74
64.75	9	7.19	8	10.31	82.44	34.92
65	9	7.22	8	10.28	82.22	35.10
65.25	9	7.25	8	10.25	82.00	35.27
65.5	9	7.28	8	10.22	81.78	35.45
65.75	9	7.31	8	10.19	81.56	35.63
66	9	7.33	8	10.17	81.33	35.80
66.25	9	7.36	8	10.14	81.11	35.98
66.5	9	7.39	8	10.11	80.89	36.16
66.75	9	7.42	8	10.08	80.67	36.34
67	9	7.44	8	10.06	80.44	36.51
67.25	9	7.47	8	10.03	80.22	36.69
67.5	9	7.50	8	10.00	80.00	36.87
67.75	9	7.53	8	9.97	79.78	37.05
68	9	7.56	8	9.94	79.56	37.23
68.25	9	7.58	8	9.92	79.33	37.41
68.5	9	7.61	8	9.89	79.11	37.58
68.75	9	7.64	8	9.86	78.89	37.76
69	10	6.90	9	10.60	95.40	33.06
69.25	10	6.93	9	10.58	95.18	33.22
69.5	10	6.95	9	10.55	94.95	33.38
69.75	10	6.98	9	10.53	94.73	33.53
70	10	7.00	9	10.50	94.50	33.69
70.25	10	7.03	9	10.48	94.28	33.85
70.5	10	7.05	9	10.45	94.05	34.01
70.75	10	7.08	9	10.43	93.83	34.16
71	10	7.10	9	10.40	93.60	34.32
71.25	10	7.13	9	10.38	93.38	34.48
71.5	10	7.15	9	10.35	93.15	34.64
71.75	10	7.18	9	10.33	92.93	34.80
72	10	7.20	9	10.30	92.70	34.95
72.25	10	7.23	9	10.28	92.48	35.11
72.5	10	7.25	9	10.25	92.25	35.27
72.75	10	7.28	9	10.23	92.03	35.43

Total Rise	Risers	Riser Height	Steps	Tread	Total Run	Angle
73	10	7.30	9	10.20	91.80	35.59
73.25	10	7.33	9	10.18	91.58	35.75
73.5	10	7.35	9	10.15	91.35	35.91
73.75	10	7.38	9	10.13	91.13	36.07
74	10	7.40	9	10.10	90.90	36.23
74.25	10	7.43	9	10.08	90.68	36.39
74.5	10	7.45	9	10.05	90.45	36.55
74.75	10	7.48	9	10.03	90.23	36.71
75	10	7.50	9	10.00	90.00	36.87
75.25	10	7.53	9	9.98	89.78	37.03
75.5	10	7.55	9	9.95	89.55	37.19
75.75	10	7.58	9	9.93	89.33	37.35
76	10	7.60	9	9.90	89.10	37.51
76.25	10	7.63	9	9.88	88.88	37.67
76.5	11	6.95	10	10.55	105.45	33.40
76.75	11	6.98	10	10.52	105.23	33.55
77	11	7.00	10	10.50	105.00	33.69
77.25	11	7.02	10	10.48	104.77	33.83
77.5	11	7.05	10	10.45	104.55	33.98
77.75	11	7.07	10	10.43	104.32	34.12
78	11	7.09	10	10.41	104.09	34.26
78.25	11	7.11	10	10.39	103.86	34.41
78.5	11	7.14	10	10.36	103.64	34.55
78.75	11	7.16	10	10.34	103.41	34.70
79	11	7.18	10	10.32	103.18	34.84
79.25	11	7.20	10	10.30	102.95	34.98
79.5	11	7.23	10	10.27	102.73	35.13
79.75	11	7.25	10	10.25	102.50	35.27
80	11	7.27	10	10.23	102.27	35.42
80.25	11	7.30	10	10.20	102.05	35.56
80.5	11	7.32	10	10.18	101.82	35.71
80.75	11	7.34	10	10.16	101.59	35.85
81	11	7.36	10	10.14	101.36	36.00
81.25	11	7.39	10	10.11	101.14	36.14
81.5	11	7.41	10	10.09	100.91	36.29
81.75	11	7.43	10	10.07	100.68	36.43
82	11	7.45	10	10.05	100.45	36.58
82.25	11	7.48	10	10.02	100.23	36.72
82.5	11	7.50	10	10.00	100.00	36.87

Total Rise	Risers	Riser Height	Steps	Tread	Total Run	Angle
82.75	11	7.52	10	9.98	99.77	37.02
83	11	7.55	10	9.95	99.55	37.16
83.25	11	7.57	10	9.93	99.32	37.31
83.5	11	7.59	10	9.91	99.09	37.45
83.75	11	7.61	10	9.89	98.86	37.60
84	11	7.64	10	9.86	98.64	37.75
84.25	12	7.02	11	10.48	115.27	33.82
84.5	12	7.04	11	10.46	115.04	33.95
84.75	12	7.06	11	10.44	114.81	34.08
85	12	7.08	11	10.42	114.58	34.22
85.25	12	7.10	11	10.40	114.35	34.35
85.5	12	7.13	11	10.38	114.13	34.48
85.75	12	7.15	11	10.35	113.90	34.61
86	12	7.17	11	10.33	113.67	34.74
86.25	12	7.19	11	10.31	113.44	34.88
86.5	12	7.21	11	10.29	113.21	35.01
86.75	12	7.23	11	10.27	112.98	35.14
87	12	7.25	11	10.25	112.75	35.27
87.25	12	7.27	11	10.23	112.52	35.40
87.5	12	7.29	11	10.21	112.29	35.54
87.75	12	7.31	11	10.19	112.06	35.67
88	12	7.33	11	10.17	111.83	35.80
88.25	12	7.35	11	10.15	111.60	35.94
88.5	12	7.38	11	10.13	111.38	36.07
88.75	12	7.40	11	10.10	111.15	36.20
89	12	7.42	11	10.08	110.92	36.34
89.25	12	7.44	11	10.06	110.69	36.47
89.5	12	7.46	11	10.04	110.46	36.60
89.75	12	7.48	11	10.02	110.23	36.74
90	12	7.50	11	10.00	110.00	36.87
90.25	12	7.52	11	9.98	109.77	37.00
90.5	12	7.54	11	9.96	109.54	37.14
90.75	12	7.56	11	9.94	109.31	37.27
91	12	7.58	11	9.92	109.08	37.41
91.25	13	7.02	12	10.48	125.77	33.81
91.5	13	7.04	12	10.46	125.54	33.93
91.75	13	7.06	12	10.44	125.31	34.05
92	13	7.08	12	10.42	125.08	34.18
92.25	13	7.10	12	10.40	124.85	34.30

Total Rise	Risers	Riser Height	Steps	Tread	Total Run	Angle
92.5	13	7.12	12	10.38	124.62	34.42
92.75	13	7.13	12	10.37	124.38	34.54
93	13	7.15	12	10.35	124.15	34.66
93.25	13	7.17	12	10.33	123.92	34.78
93.5	13	7.19	12	10.31	123.69	34.91
93.75	13	7.21	12	10.29	123.46	35.03
94	13	7.23	12	10.27	123.23	35.15
94.25	13	7.25	12	10.25	123.00	35.27
94.5	13	7.27	12	10.23	122.77	35.39
94.75	13	7.29	12	10.21	122.54	35.52
95	13	7.31	12	10.19	122.31	35.64
95.25	13	7.33	12	10.17	122.08	35.76
95.5	13	7.35	12	10.15	121.85	35.89
95.75	13	7.37	12	10.13	121.62	36.01
96	13	7.38	12	10.12	121.38	36.13
96.25	13	7.40	12	10.10	121.15	36.25
96.5	13	7.42	12	10.08	120.92	36.38
96.75	13	7.44	12	10.06	120.69	36.50
97	13	7.46	12	10.04	120.46	36.62
97.25	13	7.48	12	10.02	120.23	36.75
97.5	13	7.50	12	10.00	120.00	36.87
97.75	13	7.52	12	9.98	119.77	36.99
98	14	7.00	13	10.50	136.50	33.69
98.25	14	7.02	13	10.48	136.27	33.80
98.5	14	7.04	13	10.46	136.04	33.92
98.75	14	7.05	13	10.45	135.80	34.03
99	14	7.07	13	10.43	135.57	34.14
99.25	14	7.09	13	10.41	135.34	34.25
99.5	14	7.11	13	10.39	135.11	34.37
99.75	14	7.13	13	10.38	134.88	34.48
100	14	7.14	13	10.36	134.64	34.59
100.25	14	7.16	13	10.34	134.41	34.71
100.5	14	7.18	13	10.32	134.18	34.82
100.75	14	7.20	13	10.30	133.95	34.93
101	14	7.21	13	10.29	133.71	35.05
101.25	14	7.23	13	10.27	133.48	35.16
101.5	14	7.25	13	10.25	133.25	35.27
101.75	14	7.27	13	10.23	133.02	35.39
102	14	7.29	13	10.21	132.79	35.50

Total Rise	Risers	Riser Height	Steps	Tread	Total Run	Angle
102.25	14	7.30	13	10.20	132.55	35.61
102.5	14	7.32	13	10.18	132.32	35.73
102.75	14	7.34	13	10.16	132.09	35.84
103	14	7.36	13	10.14	131.86	35.96
103.25	14	7.38	13	10.13	131.63	36.07
103.5	14	7.39	13	10.11	131.39	36.18
103.75	14	7.41	13	10.09	131.16	36.30
104	14	7.43	13	10.07	130.93	36.41
104.25	14	7.45	13	10.05	130.70	36.53
104.5	14	7.46	13	10.04	130.46	36.64
104.75	14	7.48	13	10.02	130.23	36.76
105	14	7.50	13	10.00	130.00	36.87
105.25	15	7.02	14	10.48	146.77	33.80
105.5	15	7.03	14	10.47	146.53	33.90
105.75	15	7.05	14	10.45	146.30	34.01
106	15	7.07	14	10.43	146.07	34.11
106.25	15	7.08	14	10.42	145.83	34.22
106.5	15	7.10	14	10.40	145.60	34.32
106.75	15	7.12	14	10.38	145.37	34.43
107	15	7.13	14	10.37	145.13	34.53
107.25	15	7.15	14	10.35	144.90	34.64
107.5	15	7.17	14	10.33	144.67	34.74
107.75	15	7.18	14	10.32	144.43	34.85
108	15	7.20	14	10.30	144.20	34.95
108.25	15	7.22	14	10.28	143.97	35.06
108.5	15	7.23	14	10.27	143.73	35.17
108.75	15	7.25	14	10.25	143.50	35.27
109	15	7.27	14	10.23	143.27	35.38
109.25	15	7.28	14	10.22	143.03	35.48
109.5	15	7.30	14	10.20	142.80	35.59
109.75	15	7.32	14	10.18	142.57	35.70
110	15	7.33	14	10.17	142.33	35.80
110.25	15	7.35	14	10.15	142.10	35.91
110.5	15	7.37	14	10.13	141.87	36.02
110.75	15	7.38	14	10.12	141.63	36.12
111	15	7.40	14	10.10	141.40	36.23
111.25	15	7.42	14	10.08	141.17	36.34
111.5	15	7.43	14	10.07	140.93	36.44
111.75	15	7.45	14	10.05	140.70	36.55

Total Rise	Risers	Riser Height	Steps	Tread	Total Run	Angle
112	15	7.47	14	10.03	140.47	36.66
112.25	15	7.48	14	10.02	140.23	36.76
112.5	15	7.50	14	10.00	140.00	36.87
112.75	16	7.05	15	10.45	156.80	33.99
113	16	7.06	15	10.44	156.56	34.08
113.25	16	7.08	15	10.42	156.33	34.18
113.5	16	7.09	15	10.41	156.09	34.28
113.75	16	7.11	15	10.39	155.86	34.38
114	16	7.13	15	10.38	155.63	34.48
114.25	16	7.14	15	10.36	155.39	34.58
114.5	16	7.16	15	10.34	155.16	34.68
114.75	16	7.17	15	10.33	154.92	34.78
115	16	7.19	15	10.31	154.69	34.88
115.25	16	7.20	15	10.30	154.45	34.97
115.5	16	7.22	15	10.28	154.22	35.07
115.75	16	7.23	15	10.27	153.98	35.17
116	16	7.25	15	10.25	153.75	35.27
116.25	16	7.27	15	10.23	153.52	35.37
116.5	16	7.28	15	10.22	153.28	35.47
116.75	16	7.30	15	10.20	153.05	35.57
117	16	7.31	15	10.19	152.81	35.67
117.25	16	7.33	15	10.17	152.58	35.77
117.5	16	7.34	15	10.16	152.34	35.87
117.75	16	7.36	15	10.14	152.11	35.97
118	16	7.38	15	10.13	151.88	36.07
118.25	16	7.39	15	10.11	151.64	36.17
118.5	16	7.41	15	10.09	151.41	36.27
118.75	16	7.42	15	10.08	151.17	36.37
119	16	7.44	15	10.06	150.94	36.47
119.25	16	7.45	15	10.05	150.70	36.57
119.5	16	7.47	15	10.03	150.47	36.67
119.75	16	7.48	15	10.02	150.23	36.77
120	16	7.50	15	10.00	150.00	36.87
120.25	17	7.07	16	10.43	166.82	34.15
120.5	17	7.09	16	10.41	166.59	34.25
120.75	17	7.10	16	10.40	166.35	34.34
121	17	7.12	16	10.38	166.12	34.43
121.25	17	7.13	16	10.37	165.88	34.53
121.5	17	7.15	16	10.35	165.65	34.62

Total Rise	Risers	Riser Height	Steps	Tread	Total Run	Angle
121.75	17	7.16	16	10.34	165.41	34.71
122	17	7.18	16	10.32	165.18	34.81
122.25	17	7.19	16	10.31	164.94	34.90
122.5	17	7.21	16	10.29	164.71	34.99
122.75	17	7.22	16	10.28	164.47	35.09
123	17	7.24	16	10.26	164.24	35.18
123.25	17	7.25	16	10.25	164.00	35.27
123.5	17	7.26	16	10.24	163.76	35.37
123.75	17	7.28	16	10.22	163.53	35.46
124	17	7.29	16	10.21	163.29	35.55
124.25	17	7.31	16	10.19	163.06	35.65
124.5	17	7.32	16	10.18	162.82	35.74
124.75	17	7.34	16	10.16	162.59	35.83
125	17	7.35	16	10.15	162.35	35.93
125.25	17	7.37	16	10.13	162.12	36.02
125.5	17	7.38	16	10.12	161.88	36.12
125.75	17	7.40	16	10.10	161.65	36.21
126	17	7.41	16	10.09	161.41	36.30
126.25	17	7.43	16	10.07	161.18	36.40
126.5	17	7.44	16	10.06	160.94	36.49
126.75	17	7.46	16	10.04	160.71	36.59
127	17	7.47	16	10.03	160.47	36.68
127.25	17	7.49	16	10.01	160.24	36.78
127.5	17	7.50	16	10.00	160.00	36.87
127.75	18	7.10	17	10.40	176.85	34.30
128	18	7.11	17	10.39	176.61	34.39
128.25	18	7.13	17	10.38	176.38	34.48
128.5	18	7.14	17	10.36	176.14	34.57
128.75	18	7.15	17	10.35	175.90	34.66
129	18	7.17	17	10.33	175.67	34.74
129.25	18	7.18	17	10.32	175.43	34.83
129.5	18	7.19	17	10.31	175.19	34.92
129.75	18	7.21	17	10.29	174.96	35.01
130	18	7.22	17	10.28	174.72	35.10
130.25	18	7.24	17	10.26	174.49	35.18
130.5	18	7.25	17	10.25	174.25	35.27
130.75	18	7.26	17	10.24	174.01	35.36
131	18	7.28	17	10.22	173.78	35.45
131.25	18	7.29	17	10.21	173.54	35.54

Total Rise	Risers	Riser Height	Steps	Tread	Total Run	Angle
131.5	18	7.31	17	10.19	173.31	35.63
131.75	18	7.32	17	10.18	173.07	35.71
132	18	7.33	17	10.17	172.83	35.80
132.25	18	7.35	17	10.15	172.60	35.89
132.5	18	7.36	17	10.14	172.36	35.98
132.75	18	7.38	17	10.13	172.13	36.07
133	18	7.39	17	10.11	171.89	36.16
133.25	18	7.40	17	10.10	171.65	36.25
133.5	18	7.42	17	10.08	171.42	36.34
133.75	18	7.43	17	10.07	171.18	36.42
134	18	7.44	17	10.06	170.94	36.51
134.25	18	7.46	17	10.04	170.71	36.60
134.5	18	7.47	17	10.03	170.47	36.69
134.75	18	7.49	17	10.01	170.24	36.78
135	18	7.50	17	10.00	170.00	36.87
135.25	19	7.12	18	10.38	186.87	34.44
135.5	19	7.13	18	10.37	186.63	34.52
135.75	19	7.14	18	10.36	186.39	34.60
135	19	7.11	18	10.39	187.11	34.35
135.25	19	7.12	18	10.38	186.87	34.44
135.5	19	7.13	18	10.37	186.63	34.52
135.75	19	7.14	18	10.36	186.39	34.60
136	19	7.16	18	10.34	186.16	34.69
136.25	19	7.17	18	10.33	185.92	34.77
136.5	19	7.18	18	10.32	185.68	34.85
136.75	19	7.20	18	10.30	185.45	34.94
137	19	7.21	18	10.29	185.21	35.02
137.25	19	7.22	18	10.28	184.97	35.11
137.5	19	7.24	18	10.26	184.74	35.19
137.75	19	7.25	18	10.25	184.50	35.27
138	19	7.26	18	10.24	184.26	35.36
138.25	19	7.28	18	10.22	184.03	35.44
138.5	19	7.29	18	10.21	183.79	35.52
138.75	19	7.30	18	10.20	183.55	35.61
139	19	7.32	18	10.18	183.32	35.69
139.25	19	7.33	18	10.17	183.08	35.78
139.5	19	7.34	18	10.16	182.84	35.86
139.75	19	7.36	18	10.14	182.61	35.94
140	19	7.37	18	10.13	182.37	36.03

Decimals to Inches Chart

One slash represents a foot (')or 12' translates into twelve feet.

Two slashes or quotation marks represent inches (") or 14" translates into fourteen inches.

These marks are usually used by architects and designers and can be found on building blueprints.

Decimal	Fraction
.0625	One sixteenth of an inch or 1/16"
.125	One eight of an inch or 1/8"
.1875	Three sixteenths of an inch or 3/16"
.25	One quarter inch or 1/4"
.3125	Five sixteenths of an inch or 5/16"
.375	Three eights of an inch or 3/8"
.4375	Seven sixteenths of an inch or 7/16"
.5	One half inch or 1/2"
.5625	Nine sixteenths of an inch or 9/16"
.625	Five eights of an inch or 5/8"
.6875	Eleven sixteenths of an inch or 11/16"
.75	Three quarters of an inch or 3/4"
.8125	Thirteen sixteenths of an inch or 13/16"
.875	Seven eights of an inch or 7/8"
.9375	Fifteen sixteenths of an inch or 15/16"
1.0	One inch or 1"

Examples of Bracket Stair Construction

The illustrations in this chapter are specifically designed to provide you with more clarity on how different stairways using brackets can be assembled.

Example 1 - 3 Step Stairway

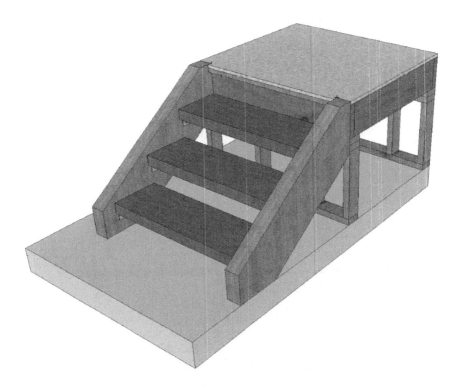

This is a simplified version of a stairway with three steps.

Bottom view of stairway.

Example 2 – Support Walls

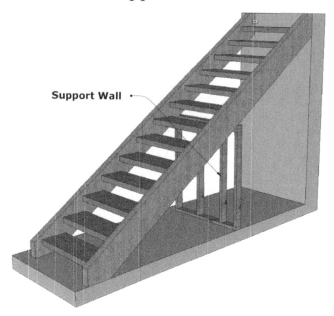

Supporting walls provide additional structural support and strengthen long stairways.

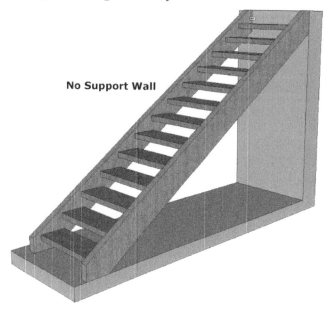

Example 3 – Stairs with Landing

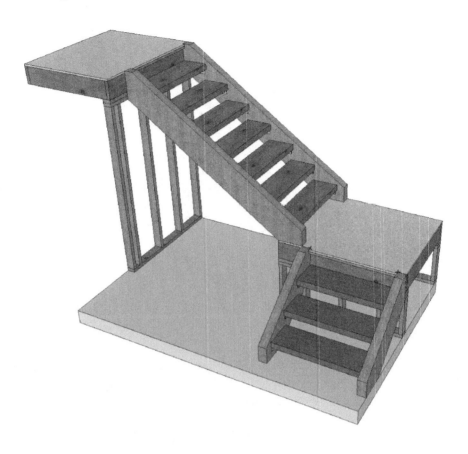

Example 3 will provide you with a few illustrations on how you can install a landing in this type of stairway.

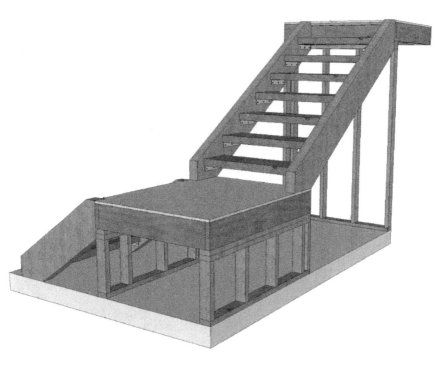
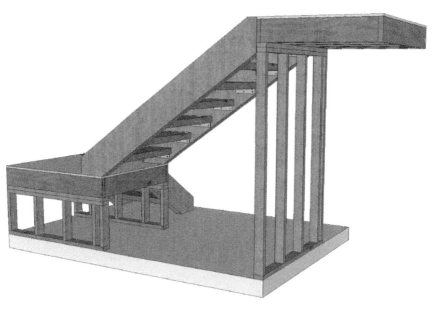

The next four illustrations will provide you with different locations for stair stringers. You might find this helpful if you're working in tight spaces.

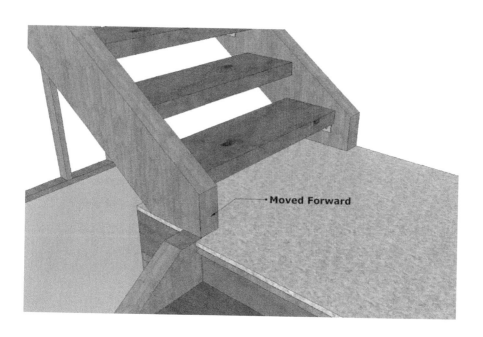
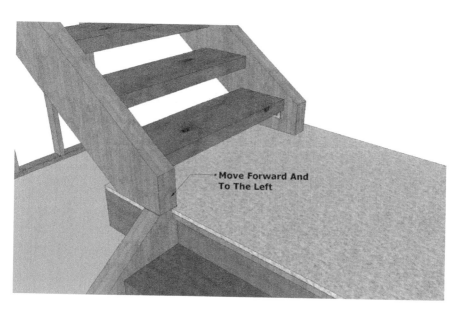

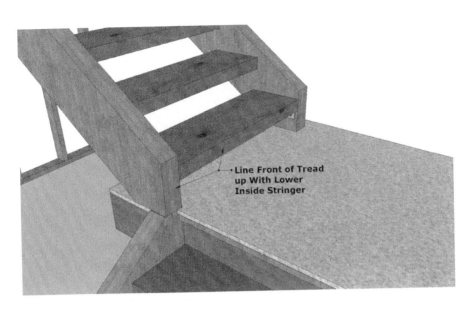

It would be hard to imagine moving the stairway any further forward, without creating safety problems.

Riser Problem and Solutions

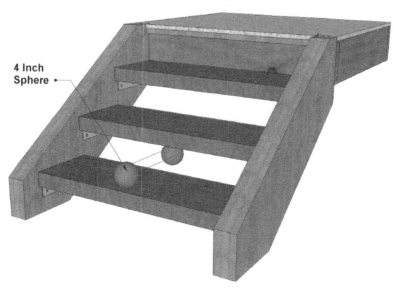

Most stair building codes require that a 4 inch sphere or round ball cannot pass through any part of the stairway. Visit our website for more information or follow one of the methods for installing a blocker board.

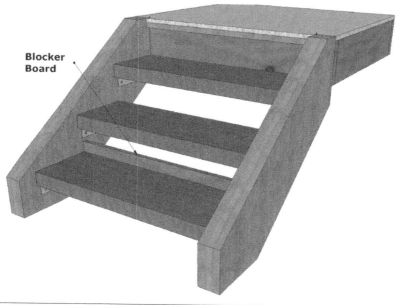

The blocker board can usually be attached with nails or screws to the back of a large stair tread. The illustration below blocks the entire riser area.

The blocker board can also be attached to the bottom of a stair tread.

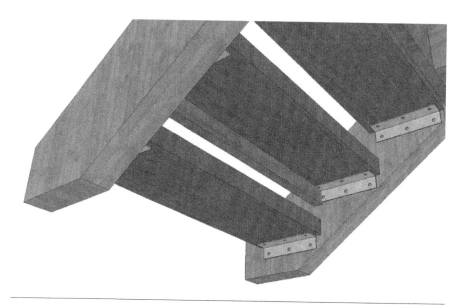

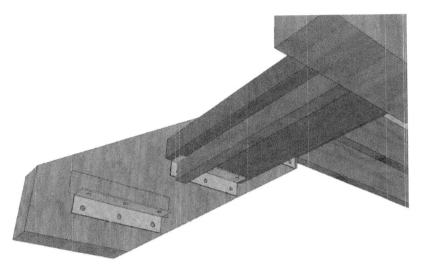

It can be moved back, just don't move it too far back, otherwise it won't block the 4 inch sphere.

Extended Stringers Can Create Safety Problems

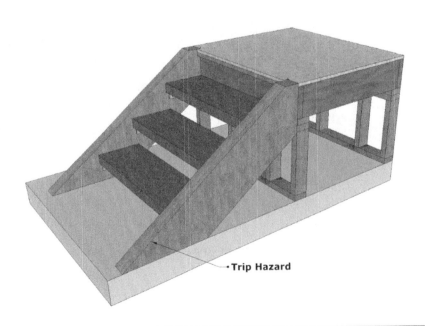

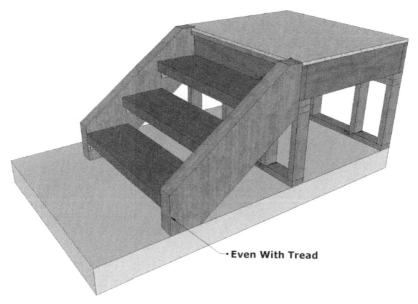

Even With Tread

The stringers can be cut even with the tread or even extend a couple of inches past the face of the bottom tread as shown in illustration below.

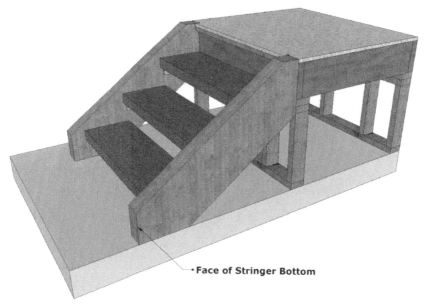

Face of Stringer Bottom

Post Base Under Stringer

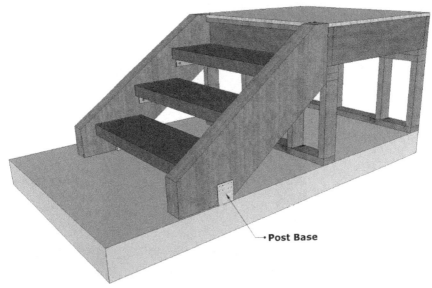

If you're building in exterior stairway then this method could prevent stringers from dry rot or water damage. Exterior wood shouldn't touch concrete, because moisture can be absorbed from the concrete into the wood.

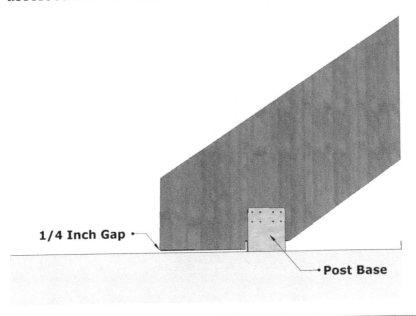

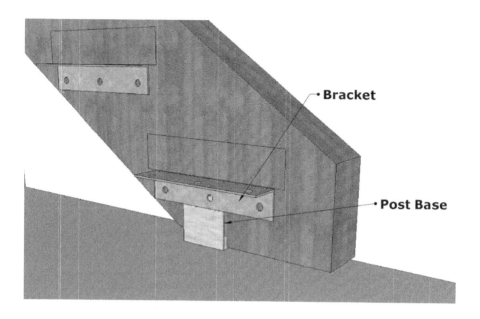

Post base can create problems for installing brackets and might need to be installed over the post base or modified.

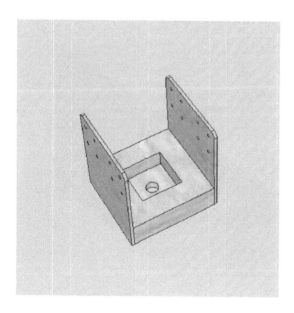

Made in the USA
Middletown, DE
24 November 2023

43404006R00076